SCOTLAND'S PAST IN ACTION

Fishing
and
Whaling

Angus Martin

N·M·S

NATIONAL MUSEUMS OF SCOTLAND

Published by the National Museums of Scotland
Chambers Street, Edinburgh EH1 1JF

ISBN 0948636 67 X

© Angus Martin and the Trustees of the National Museums of Scotland 1995

British Library Cataloguing in Publication Data
A catalogue record for this book is available for the British Library

Series editor Iseabail Macleod
Picture research Susan Irvine
Designed and produced by the Publications Office of the National Museums of Scotland
Printed by Ritchie of Edinburgh

for my daughters, Sarah, Amelia and Isabella

Acknowledgements

It has been no easy task selecting and accommodating, within a limited space, the essentials of such a huge subject. My approach, inescapably, has been subjective, so this work represents to no small degree my own interests. The reader will therefore find relatively little political and economic history, and few statistics. I have preferred to concentrate on the lives and labours of the fishing communities and the tools they used.

I have been helped by many people, to all of whom I extend my deepest gratitude. I wish particularly to mention: Bob Smith, Cauldcoats, for his invaluable assistance at all stages of the work; Iseabail Macleod and Jenni Calder, the editors of this series; Geoff Swinney of the NMS; the staff of Campbeltown Public Library; Jim, John and Nancy of the Fishery Office, Campbeltown; the library staff, Marine Laboratory, Aberdeen; Susan Irvine, picture editor; Roger Leitch; James R Nicolson, Scalloway; and my wife Judy, for guiding me through the trauma of word-processing.

Front cover: *The herring fleet goes out. A detail from* Stonehaven Harbour *by William F Douglas, 1876.*

Back cover: *From an illustration for the 'Eyemouth tapestry'.*

CONTENTS

FISHING AND WHALING

1 Food from the shore

In the history of this region, now called Scotland, there has probably never been a time when the shore was unimportant to survival. Even today there are people who make their living by scouring the ebb for shellfish, particularly winkles. The middens of our ancestors testify to the importance of shellfish-gathering. At MacArthur's Cave, Oban - a rock-shelter in use as far back as the mesolithic period, some 5,000 years ago - an excavated midden contained a heap of shells, including limpets, winkles, whelks, mussels, oysters and scallops. In extensive midden deposits at Inveravon and Polmonthill in the Forth Valley, an abundance of shells of oyster, winkle, mussel, cockle and whelk was left behind by these migratory mesolithic tribes, who neither kept livestock nor planted seed.

The shore in later times was especially crucial to survival in years of famine. Crops might fail and livestock perish, but there was always shellfish for the taking when the tide went out. Oatmeal, the mainstay of the people's diet, generally began to run out in spring, and from then until crops were harvested the shore was the chief resort of the common people.

Cockles, which are found in sandy bays and strands, were gathered in great quantities. Of the parish of Mid and South Yell in the Shetlands, it was written, in 1841, that ' ... often, in times of scarcity, (the cockle) has been the means of saving the lives of hundreds of inhabitants'. From the western strands of the Outer Hebrides, vast quantities were gathered annually. In two bad summers in the late eighteenth century, from 100 to 200 horse-

For thousands of years Scots have gathered food from the sea and shore. Discharging sprats at Tarbert, Loch Fyne, December 1989 from the Antares *of Carradale.* Lachlan Paterson

Traigh Mhór, Isle of Barra, a prolific cockle-shore.
Industry continues there: the raked cockles are sold to local
agents for export to the Continent. SEA

loads of cockles were carried off the Barra sands alone at low
water every day of the spring tides. Hundreds of islanders would
congregate to dig out the cockles and load them into creels.

These cockles could be boiled and eaten straight out of the
shells or else stewed in milk, if milk was available, and eaten as a
soup. Winkles too were turned into soup, or else, in later and
more prosperous times, eaten in white sauce. Winkles in particular were taken medicinally in spring, to 'purify the blood'.

The common limpet features prominently in midden deposits
from earliest times. The explanation is not hard to reach. Though
tough and unappetizing, at least to modern taste, limpets are
sure prey. They can be found without the least difficulty, stuck to
the rocks at almost all stages of the tide, even on the most exposed
of Atlantic coastlines. There is, however, a knack to dislodging
them, because when disturbed in the least they clamp firmly to
the rock. Slender pebbles, usually chipped at one or both ends,
found in the mesolithic shell-middens of Oronsay in the Inner
Hebrides, are now believed to have been limpet-hammers, for
knocking limpets off the rocks. Neil Gunn's classic novel of the

Caithness herring fishery, *The Silver Darlings*, opens with Tormad, father of the protagonist, Finn, trying to 'flick a limpet out of the boiling pot', and burning his fingers and upsetting the pot over the fire.

The sheer abundance and durability of shell deposits have perhaps tended to overshadow the importance of fish in the diet of the earliest peoples. But the evidence is there. In King's Cave, Jura, identified fish bones include several species, such as haddock, ling and hake, which could not have been taken but with hook and line; and, again, the evidence is there, in the bone fishhooks found, for example, on Oronsay and on the islet of Risga in Loch Sunart, Argyll. The fish bones themselves, in times of scarcity, could be pounded and used to eke out meal.

Most shellfish are sedentary and easily taken; fish proper are less easily taken. Yet opportunities would arise. The small silvery sand-eel, which sometimes buries itself in the sand during ebb tide, could be flicked out using purposely-made iron hooks or old sickles. The spearing of fish was common practice around the coasts, using an iron fork, the shaft of which could extend to 25ft (7.62m). In the Solway Firth, in the eighteenth century, the spearing could be done from horseback, the spear killing 'at a

A flounder speared. The man has a bag slung to the back, for holding his catch. West Coast, date and location unknown.
SEA

great distance'. And flatfish could be speared with nothing more sophisticated than a sharpened stick!

Fish could occasionally be collected on the shore - cast up by storms or chased out of the water by marine predators - or from rock-pools in which they had become trapped with the falling of the tide. That circumstance no doubt led, in ancient times, to the notion of the *yair* (Scots) or *cairidh* (Gaelic). If fish could be trapped in nature, why not improve on nature? The yair was basically an enclosure of wattle or stone, built in or across a bay or the head of a loch or estuary, and it operated on a very simple principle. The fish came in over it when the tide was flooding, and were caught behind it when the tide ebbed. Herring were the chief objects of the yair-fishermen in many places, but all other kinds of shallow-water fish could be taken.

In the parish of Cardross, on the Clyde estuary, there were many such yairs in the eighteenth century, but by 1839 only two were still maintained, at Ardmore and Colgrain. Catches were generally removed from the yair using lap-nets, small bags attached to poles, but at times a yair would hold so much herring as to defeat the resources of the local population. These fish-traps could therefore be very wasteful and were condemned too for their indiscriminate destruction of immature fish.

The immense importance of the coalfish or saithe, since earliest times, is demonstrated by the profusion of names it has had in its various stages of growth. These clusters of names varied from locality to locality. The saithe was abundant on all coasts, and there was no other fish on which the people relied so heavily. Not only was the saithe plentiful, but it was also dependable, unlike the herring which was notorious for its erratic movements.

As there were many names for the fish, so too there were many ways of catching them. With rod and line from boats, rocks or quays was as popular a method as any, and had its recreational value too. Certain rocks on the shore, suitable for casting from, became traditional *craigseats*. Some, indeed, came to bear the names of particular men with whom they were especially associated. The customary bait was limpets, some of which might first be pulverized and cast into the sea to draw fish to the spot. *Pock-nets* were also used from the rocks to catch saithe, but were more expensive and time-consuming to make. Basically, these were bags - a pock or poke in Scots is simply a bag - attached to iron hoops, from about 4ft to 10ft (1.22-3.05m) in diameter, and suspended from a long pole. The bag was baited and weighted with a stone to sink it, and was capable of taking as much as half-a-hundredweight of fish.

Screenges, small beach-seines, hauled repeatedly where saithe were known to be shoaling, sometimes took great quantities. At

Working a pock-net *for saithe from the rocks, Shetland. It is unlikely that the boy with rod and line would have been casting so close to the net-fisher in reality.* SEA

9

Kirkmaiden in Wigtownshire, in the early nineteenth century, fifteen cart-loads could be caught 'during a single tide'. In Lewis, at that same period, a net was improvised by sewing half-a-dozen blankets together and dragging one end of the assemblage to the shore at the shallow mouth of a burn.

The flesh of the young saithe was esteemed in some communities and despised in others, probably owing to its connotations with poverty and famine. But, particularly in the Highlands and Islands and the Northern Isles, the young coalfish formed an indispensable element in the diet. They would be eaten fresh or hung out to dry in the sun for winter provisions. These fish also supplied liver-oil, extracted by slow heating, a benefit which

Sun-drying saithe strung around the fence of Archibald MacCormick's thatched house at Loch Eynort, South Uist, 1920s. Margaret Fay Shaw Campbell, SEA

cannot be over-stressed. Oil was essential in the treatment of leather, for example, but its main value was as fuel for the *cruisie*, the metal lamp, with a wick of plaited rushes, which was in universal use on the coasts and islands. The cruisie did not give off much light - about the equivalent of one candle - but it was all that the people had.

Surplus oil could be sold, and it was said of Barra in the late eighteenth century that some of the islanders had 'even been known to pay their rents with the oil extracted from the small fish called *cuddy*'. In good seasons, great quantities of oil were produced from the humble coalfish, an estimated 2,000-plus barrels, for example, in Shetland, from October 1790 to April 1791.

2 Dogfish and sharks

There are, in Scottish waters, several small shark species known as dogfish, of which the piked dogfish or spurdog was of greatest economic value. Dogfish move in vast voracious shoals and were hated by fishermen owing to their destructiveness to nets and to their habit of taking bites out of fish caught in drift-nets and on lines. Dogfish, however, produced fine oil, the livers of about 20 fish yielding a Scotch pint (or four Imperial pints). Using handlines, one man could take 300 dogfish in a day, and on some parts of the coast dogfish constituted a specific fishery. In August, the fishermen of Nigg in Aberdeenshire pursued them using a special *dog-line*, which was of stronger materials than the ordinary line and carried bigger hooks.

The tenants of Ness, Lewis, were reputed, in the early nineteenth century, to be paying their rents 'by dogfish oil alone', and in the late eighteenth century the Rev Mr Liddell in Orkney was writing: 'When the dog-fishing fails, which sometimes happens, the people are in the utmost distress from want of oil, which then rises from 6d or 8d per Scotch pint, to 1s or even 1s 6d.'

Spurdogs, like coalfish, were eaten in some parts of the country and disregarded in others. They could be dried or cured, and sometimes cured and then smoked. Largely, however, they were eaten only by poor folk. The carcasses, after extraction of the livers, were more likely to end up putrefying on dung-heaps to fertilize the farmers' fields.

When the King's Cave on Jura was excavated in the 1970s, among the organic artefacts were found fourteen spurdog spines. These spines, or spurs, from which the species takes its name, protrude immediately in front of both dorsal fins. One of them had been worked into a tool, which showed signs of use. A similar spine from a site in Denmark, dated 3,200 to 3,000 BC, was apparently used as a combined awl and needle.

The lesser spotted dogfish was also eaten, and considered 'very delicate', according to a report from Jura and Colonsay in the late eighteenth century. The Kintyre fishermen formerly used its rough skin for scouring off varnish and paint when the boats were beached for spring-cleaning.

That monstrous relative of the dogfish, the basking shark, was also hunted for its liver-oil. Its local names were *sail-fish*, after its tall, sail-like dorsal fin; *sun-fish*; *muldoan* and, in Gaelic, *cearban*. It too was hated by fishermen, owing to the damage its huge bulk caused to nets. In the summer of 1922, off Barra, drift-netters were each losing up to 20 nets in a night to basking sharks. Not only did the basking shark tear nets, but its slime coated the nets, and the affected parts had to be cut out and replaced, otherwise they rotted. The basking shark is, after the whale shark of the Pacific, the biggest fish in the world. It can approach 30ft (9.14m) in length and weigh six tons, of which the liver alone can account for one ton. Surprisingly little is known about its natural history, particularly its migrations and its breeding habits, for it disappears into deep water in autumn when the planktonic life, on which it feeds, also disappears.

Given its prodigious oil-yield, it is hardly surprising that the basking shark was hunted widely on the western coasts of Ireland and Scotland. Written sources for the eighteenth and nineteenth centuries attest to successful fisheries from Arran, Ayrshire, Coll,

Tiree, Barra, South Uist, and no doubt elsewhere. The English naturalist and traveller, Thomas Pennant, has left a vivid account of shark-fishing from Arran in the late eighteenth century: the harpooning of the fish and their plunge to the seabed, where they frequently coiled the rope around themselves by 'rolling on the ground'; their immense strength in towing boats for twelve and sometimes twenty-four hours before succumbing and being hauled ashore or to the boat's side for extraction of the liver, 'the only useful part'.

The fishery began to decline in the early nineteenth century, owing to the withdrawal of the Board of Fisheries bounty on shark oil, a valuable inducement; and the coming of the paraffin lamp, in the mid-nineteenth century, effectively ended the reliance on fish oil.

Shark-fishing was revived in Scottish waters in 1938 by Anthony Watkins, operating from Carradale in Kintyre, where he built a factory to process his catches. In 1946, encouraged by the promise of high oil prices in the aftermath of war, he resumed

Harpooning a basking shark at the mouth of Loch Ranza,
Arran. From Thomas Pennant's A Tour in Scotland and
Voyage to the Hebrides, *1772.*

operations with three converted fishing vessels as harpoon boats and a converted steam-drifter as a floating factory. He had, however, a rival by then. Gavin Maxwell, the naturalist and traveller, had bought the island of Soay, off Skye, built a factory there and fitted out two catchers. Neither operation was entirely successful. By 1949 Maxwell's company was in trouble and he gave up. Watkins held out for several years more. Yet another revival of shark-fishing took place in the late 1970s, when an Ayrshire fisherman, Howard McCrindle, took it up as a seasonal alternative to trawling. That enterprise ended in 1993 with the breaking up of his vessel, the *Star of David*, under the government's 'decommissioning' scheme. Increasing concern about overfishing will probably lead to the protection of the basking shark.

3 Line-fishing

The main 'round fish' species are haddock, cod, whiting, ling and hake. Flatfish species include plaice, flounder, witch, megrim, dab, brill, sole, turbot, and halibut, which is the king of the flatfish and reaches huge proportions. The winged and sinister-looking skate and ray species, though flat, are related to the sharks.

White-fish, traditionally, were caught with baited lines. Originally, these would have been simply a hand-held line or rod and line worked from the shore or from small boats close to the shore, to supply domestic needs. But as commercial fishing caught on, long-lines were evolved to maximize catching potential. The long-line is simply an extension - miles long - of the hand line or rod and line. The essential components - hook, line and sinker - are the same, but the extent of the long line is limited only by the size of the boat and the number of crew the boat can carry.

Each crewman was responsible for his own line. He and his family had to prepare it for the fishing season, maintain it in working order, gather bait for it, process and attach the bait to the hooks, clear the lines after a day's fishing ... and then repeat the whole operation daily throughout the season. It was a tedious and

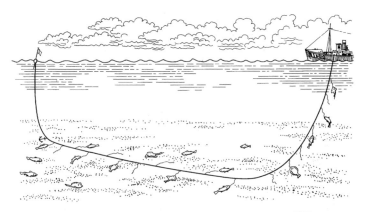

A deep-sea long-liner hauling in gear. Tim Smith, based
on a diagram in *Fish and Fisheries* by Alister Hardy

Details of long-line, showing components.
Tim Smith, based on a diagram in *An Introduction to Commercial
Fishing Gear and Methods* by E S Strange

time-consuming labour, but the great economic
benefit of line-fishing was that it provided a sure
income, however small, and a regular supply of fresh
fish for family consumption.

There were two kinds of long-line, the *sma* (small)
and the *big* or *great*. The working principles were the
same, the only differences being in scale and in the
fact that small lines were baited ashore and big lines
baited at sea. Small lines were designed for inshore
waters and caught mainly the smaller species of demersal fish
(whiting, codling, haddock, plaice, etc), while the big lines were
set in deeper waters to take ling, skate, big cod, turbot and
halibut. A typical small line, maintained by one man, would com-
prise 600 to 1,000 hooks, attached by *snuids*, a yard (0.91m)
apart, to the main line, or *back*. The snuid was part line and part
plaited horsehair, the *tippin*, to which the hook was bound with
thread. The joined lines were set between stones or anchors,

Shelling mussels for bait, Auchmithie, about 1900. SEA

linked to a marker buoy. The big line was, of course, on a bigger scale. The hooks were bigger, and the snuids spaced at wider intervals, up to five fathoms (9.14m) on the back.

A regular supply of bait is essential in line-fishing, and the varieties of bait were numerous. At small lines, mussels were preferred, and on coasts where these shellfish were scarce, the fishermen had to import them. The Morayshire line-fishermen got theirs from Tain, at £2 per boat-load in the early nineteenth century. The estuary of the River Ythan in Aberdeenshire was another major source of mussels, and a charge was levied there too: £3 yearly to young fishermen, and £2 to those above 60 years. When mussels were particularly scarce on the East Coast, supplies might be shipped in from as far off as Ireland and Holland.

Mussels, once obtained, were bedded in a *scaup* (stone enclosure) in the ebb, for convenient gathering. The gathering and preparing of bait was mainly the work of women and children.

The mussels had to be shelled by cracking them open with a short-bladed knife and scooping out the meats. If the shells were barnacle-encrusted, cloths would be wrapped around the fingers for protection.

William Smith of Cellardyke, Fife, as the oldest child in the family, would rise with his mother about four o' clock on winter mornings to do the shelling. 'I have seen her,' he wrote, 'when she had a baby in the cradle, with the cradle-string tied to her foot rocking the cradle and with her hands baiting the line.' When mussels were scarce, he had to gather limpets for bait. He also went 'up the country' to gather grass by the roadside and in fields. When withered, the grass was used for separating the layers of baited hooks in the line-basket. The hooks of the big lines required more substantial bait, such as herring, haddock, squid, and the large whelk, which was caught in creels baited with the heads and guts of fish.

Line-fishing was carried on all around the coasts. Some communities, however, specialized in it. One such was the Shetland

A line-fisherman of Scalloway, Shetland, baiting big-line hooks before going to sea. SEA

Line fishing tackle in the collections of the National Museums of Scotland.

haaf (ocean) fishermen. Many crews worked from remote fishing stations (p66-7) to which they removed in May, after the crops had been sown and the supply of peats cut. They remained there until mid-August, returning home only on occasional weekends with fish for their families who in the meantime attended to the work of the crofts. These hardy Shetlanders were accustomed to rowing their *sixerns* (six-oared open boats) up to 50 miles from land, an effort of some twelve hours' duration. About three hours after their lines had been shot, the hauling would begin. Working 1,200 hooks, 400 ling was reckoned a very good catch. At the stations, the fish were split, boned and cleaned, then salted and dried in the sun, an involved and time-consuming process.

The Shetland fishermen were not unique in the great distances they ventured from land. In the mid-nineteenth century, crews from Cullen in Banffshire fished cod, skate and ling up to 60 miles offshore, remaining at sea for 'days and nights together, in the event of the weather proving favourable'.

The main traditional markets for Scottish dried fish were Ireland and the Catholic countries of Europe, principally Spain and Portugal, for consumption during Lent and fast days. The practice of smoke-curing haddock developed in the cottages of North-East Scotland and spread south. Originally the fish, first split and then hung in the peat-smoke, were produced in small kilns or in the cottage chimneys themselves, but later the industry became factory-based. The centre of the trade was Aberdeenshire, and in and around Aberdeen itself nearly 358,000 hundredweights of smoked haddock were produced in 1903. The most famous type of 'smokie', the *Finnan haddie*, took its name from the fishing village of Findon, near Aberdeen.

Fresh fish was traditionally sold by *hawkers* or *cadgers*, usually female. Their loads, contained in *creels* (large baskets), were often carried great distances. It was a 'well-attested fact' that three fish-wives in the late eighteenth century walked from Dunbar to Edinburgh, a distance of 27 miles, each with a 200lb load on her back, in five hours; and the fish-wives of Fisherrow frequently managed one creel, in relays of three women, from there to the fish-market at Edinburgh in less than three-quarters of an hour. A lengthy and fascinating description of the lives of the fish-wives of Fisherrow can be found in the *Old Statistical Account* for the Parish of Inveresk, Midlothian. In the 1770s the poet Robert Fergusson described a Buchan, Aberdeenshire,

Drying fish, some spread and some stacked, on the rocks of Foula, Shetland, about 1902. The spread fish appear to be cod. SEA

man hawking his *speldins* (dried whitings) at Edinburgh Hallow Fair, at the end of October.

Great-lining from steam-powered vessels, chiefly from Aberdeen and Anstruther districts, gave the industry a modern form in the early 1890s, and distant-water lining in motor-boats has continued to the present time.

4 King herring

In the social and economic history of Scotland - indeed, of Europe - the herring commands an elevated status among fish. No other species has been so celebrated in folklore, music and literature. Part of the explanation lies in the herring's past importance as an abundant and cheap food. There is also, however, a mystique attached to the herring, which is less easily explained. Herring-fishing was essentially an industry of the night. The boats went out in the evening and returned in the morning, and the fishermen's work was carried on while the greater part of the population slept. That departure from land into the uncertainties of sea and darkness was, to some extent, what made herring-fishing so 'romantic'.

The hunting aspect, too, is part of the mystique. With white-fishing, the line or trawl vanishes to the sea-bottom, and is almost certain to yield a catch of some kind. But herring-fishermen had to find herring before they could catch them. If there were no herring, nights and weeks of nights might pass without a single fish being caught. And then, perhaps, the labours of one night would redeem the long wait: a big catch sold at a good price, and fantastic wealth in the pockets of the crew. Such bonanzas were what made the herring-fishing magical to fishermen themselves. A good crew was bonded by a unity of purpose, which was to hunt the herring to its death.

The herring is a beautiful fish, perfectly proportioned with a blue-green sheen on its back and a silver underbelly. The sight of a huge catch - glittering in sheets as drifts were hauled, or milling like silver on the boil in the bunt of a ring- or purse-net - excites even the most seasoned of fishermen. There is an aesthetic element to herring-fishing, which poets have not failed to catch. Hugh MacDiarmid, George Campbell Hay, Norman MacCaig, W S Graham and Naomi Mitchison have all written of the herring.

The potential of the Scottish herring fisheries was not fully realized until the nineteenth century. Until then, the Dutch had dominated the European markets with that product - pickled or cured herring - which they had brought to perfection, and which was to bring the Scots pre-eminence in their turn. Various schemes were tried to increase the competitiveness of the Scottish fishery, including a subsidized operation on the Dutch model, using *busses*, large, decked vesels from which open boats were sent out to fish with drift-nets, and aboard which the catches were cured. But the buss fishery, which began in 1750 and lasted some 80 years, was not altogether successful. The subsidies, or boun-ties, were paid out according to the tonnage of the vessels and therefore failed to stimulate the catching effort, and the operation of the busses was hampered by bureaucratic restrictions.

The real expansion of the herring fisheries did not begin until 1786, when a bounty of 2s per barrel was offered by the govern-ment for herring caught from small boats and cured according to regulations. Incentives, at last, were there for all fishermen.

Herring shoaled all around the Scottish coasts, but not always in the same places and in the same numbers. In the Firth of Forth, for example, a big winter fishery developed in the 1790s, which, at its peak, attracted from 800 to 900 boats and sustained 100 curing-yards around Burntisland. By 1805, it was over.

One of the greatest herring fisheries was centred at Wick, on the Caithness coast. By the mid-1820s, upwards of a thousand boats gathered there and at the other, smaller curing stations along the coast. At Wick, in particular, the huge population influx during the fishing season - some ten thousand persons - created real problems. Lodging-houses were packed - sometimes ten or twelve persons occupying one small room - the houses, shores and streets were filthy, 'putrescent effluvia steaming up from the fish-offals lying about everywhere', and typhoid was epidemic season after season. In 1832, 306 cases were reported in Wick alone, of which 66 proved fatal.

In the early 1900s, similar problems were being confronted in the Shetlands, where a hugely successful fishery developed. In a ten-year period from 1895 to 1904, the annual catch there shot from 449,775 hundredweights to 1,901,357 hundredweights. At one of the outlying curing stations, Balta Sound, the resident population of fewer than five hundred souls would suddenly increase, early in June, to about ten thousand, presenting serious difficulties with water and sanitation.

Many of the fishermen who crewed the herring boats were simply hired for the season. Conditions of hire varied. Either a

Wick harbour, packed with skaffies, in the 1860s.

fixed wage was paid, higher or lower, according to experience, or a share of profits paid out at the end of the season. Alternatively, the deal might be a bit of both. Fixed wages varied according to labour supply.

What were conditions like in the open boats of the nineteenth century? Geologist and historian of Cromarty, Hugh Miller:

> The profession of the herring fisherman is one of the most laborious and most exposed both to hardship and danger. From the commencement to the close of the fishing, the men who prosecute it only pass two nights of each week in bed. In all the others they sleep in open boats, with no other covering than the sail. In wet weather their hard couch proves peculiarly comfortless, and even in the most pleasant it is one upon which few besides themselves could repose... They start up on the slightest motion or noise, cast a hurried glance over the buoys of their drift, ascertaining their position with regard to the fishing bank, or to the other boats around, and then fling themselves down again.

By 1906, the great majority of Scottish fishermen were wholly dependent on herring-fishing, many having abandoned seasonal line-fishing. The immediate effect was that they were forced to spend much longer periods away from home. On the other hand, they were earning more money, and many were not only boat-owners, but home-owners as well. In early summer, they fished either the west coast of Ireland or the west coast of Scotland, from Barra Head to the Butt of Lewis. In midsummer their operations switched to the Shetlands, and in early autumn they fished the length of the east coast of Scotland, finishing their season off East Anglia.

The herring fishery of Loch Fyne was historically different from the established fisheries elsewhere in Scotland, in that its regular markets were in fresh herring, owing to the proximity of the fishing grounds to Glasgow. It appears to have been consistently productive until the present century. As far back as 1527, Hector Boece was remarking that 'in Lochfine is mair plente of hering than is in ony seis of Albion'. In the eighteenth century, in some years, upwards of 500 boats were concentrated in the narrow upper loch.

In the 1830s, however, there appeared a method of herring-fishing which was variously known as *trawling* and *seining*, but which finally came to be known as *ring-netting*. This method was pioneered by the fishermen of Tarbert, at the southern end of Loch Fyne. It immediately met with resistance from the upper loch crofter-fishermen and the curers, who complained of the destruction of herring spawn and fry and predicted the 'annihilation' of the Loch Fyne fishery. These concerns were undoubtedly genuine, but the facts were that a trawl-skiff and net represented about a quarter of the cost of even a small drifter and nets, and that the trawl-fishermen had a greater earning capacity.

The outcry against trawling was such that in 1851 the method was banned. It remained illegal until 1867, during which period the trawlers carried on defiantly. They were harassed on land and sea, jailed, and had their nets and boats confiscated. In 1853, Colin McKeich of Tarbert was shot and wounded by a naval crew, and in 1861 a young Ardrishaig fisherman, Peter McDougall, was shot dead by a marine. But after its legalization, trawling, or ring-netting, steadily gained precedence over drift-netting throughout the Clyde, and by the end of the First World War, the fishing communities of Upper Loch Fyne had virtually vanished, as indeed they had themselves predicted.

Fishermen, like farmers, work in nature. Increasingly, however, they have ceased to work with nature, and to respect certain principles to which their forefathers adhered. The destruction of spawning herring and immature fish of all species was avoided by the traditional drift-nets and long lines. Modern trawls are particularly destructive instruments, which sweep up and compress everything in their paths. Before the introduction of electronic equipment for fish-finding and for charting bad ground, and of radios which give access to weather forecasting, fishermen were attuned constantly to their environment: watching for signs of fish and for signs of change in the weather, and checking their *meiths* (marks on the land) to satisfy themselves that their gear was not drifting on to a rocky seabottom on which it could be torn or entirely destroyed.

Herring-fishermen, and particularly ring-net fishermen, preferred to shoot their nets on a sign of herring. There were many signs of herring, some of them obvious and others so elusive that only an experienced fisherman could detect them. In daylight, before fishing commenced, fishermen looked for the 'herring whale'- the collective name for common rorquals, lesser rorquals and sei whales - and for the porpoise, all of which hunted herring by a herding strategy, keeping the shoal bunched, and feeding on the outer fish. Occasionally a whale would be seen bursting to the surface clear of a dense shoal, the herring cascading in silver off

Basketing herring from the bag of a ring-net into the Busy Bee *of Campbeltown, off Heisgeir, South Minch, early 1930s.*

her back. Concentrations of gannets, striking vertically into the water, were an important sign.

Strings of bubbles rising to the surface were another sign; oily patches on the surface and the associated odour would be investigated; a red sheen on the water in winter sunlight betrayed the presence of a shoal beneath, and in dull wintry weather shoals might be seen close to the surface, swimming in 'black lumps'. Basking-sharks, which feed on the same planktonic forms as herring, might signify a potential fishing.

In late summer and autumn, when phosphorescence lit the sea, fishermen looked for herring in the night waters. The Gaelic-speaking fishermen called phosphorescence *losgadh*, in English *burning*. Sometimes shoals would be sighted, so vast in their greeny spread that fishermen spoke of 'parks' or 'fields' of herring. Real skill was required, however, to discern the dim flash of a deep-swimming shoal. On calm nights, fishermen would listen for herring. The *plowt* (plop) of one herring jumping out of the water often yielded a netful of fish, and at other times a big *play* of herring could be heard, from several miles away in quiet conditions, as a huge shoal broke the surface in a mass.

5 Nets and boats

There was much work attached to the making and maintaining of nets, and the whole process was sometimes managed by the fishermen and their families. The main home-grown net-fibre available to the Scots was flax, which was very costly to manufacture, especially when hand-spun with the drop-spindle. The spinning of flax and imported hemp and the weaving of nets had become cottage industries by the eighteenth century. At Wick, in 1841, the spinning and weaving of yarn for nets was virtually the sole occupation of females, who were paid a 'miserable pittance'.

Many fisherfolk, however, made their own nets. It was winter's work. William Smith of Cellardyke, as a boy in the mid-nineteenth century, had to weave eight rows of meshes daily when he returned from school, and sixteen rows on a Saturday. A wooden gauge was used to ensure uniformity of mesh size. Netting-needles were carved from wood, mainly elder and dog-rose, or from animal bone, which was first boiled to soften it and then flattened under a board. The whittling of needles was a pastime among fishermen.

Mechanical net-making gradually replaced weaving by hand. The first machine was patented in Britain in 1778, but it was not until James Paterson of Musselburgh introduced his design in 1820 that a new industry began to take form. By 1869, some fourteen factories, employing more than 2,000 workers, were operating in Scotland. The superior cotton had by then been introduced.

A well-built and well-maintained boat would last decades, but nets, however carefully looked after, had to be replaced frequently. Much effort was expended by the fishermen in preserving

Mending a ring-net on the New Quay, Campbeltown, about 1900. SEA

their nets from rapid decay. They had to be dried frequently, and it was the custom to spread them in fields leased for that purpose. In some communities, the women had the task of drying one set of nets while the other was in use. In the early nineteenth century, net-drying poles began to appear. These hangers, constructed of sturdy wooden uprights, usually the trunks of trees, with cross-spars for stepping on, were much more efficient and a saving on space. In some places, the stark uprights, or their stumps rotting in the ebb, are all that remain to show that a fishing community ever existed there.

Not only had nets to be dried frequently, they also had to be immersed in a protective solution to reduce bacterial damage and to protect the twine from wear and tear. Originally, the bark of trees was the only substance available, and birch was preferred because it imparted a 'softness and elasticity' to the nets. In the early nineteenth century, catechu, or cutch, distilled from the wood of acacia trees, began to be imported from the East. After the Second World War, synthetic twine entirely removed the necessity of the fishermen's having to touch their nets at all after the week's work, unless to repair damage. Damage could happen with maddening frequency, particularly when gear was being fished over rocky ground. Nets could also be damaged by dogfish, sharks and whales, and by becoming entangled with other crews' nets, and having to be cut free. A damaged net meant lost fishing time and perhaps a lost catch.

The drift-net, as a fishing instrument, changed little throughout its long history. As boats grew bigger, so too did the nets, but the simple principle held: the drift was a string, or *train*, of joined nets which hung curtain-like in the water. If herring were swimming, then the chances were some would swim into the nets and become enmeshed by the gills It was as simple as that. Often, however, the shoals were not swimming, or were swimming too deep to be taken, and the nets would remain empty. At other times, the shoals would mesh too thickly and the weight of herring would sink the nets irre-trievably. In the summer fishing off Lybster, on the Caithness

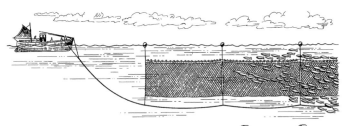

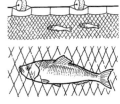

A motor drifter lying to her nets (swimming herring are not drawn to scale), and details of drift-net, showing herrings enmeshed.
Tim Smith, based on a diagram in *Fish and Fisheries* by Alister Hardy

coast, in 1886, no fewer than 1,200 drift-nets were left on the fishing grounds during the first week of August.

Most herring fishing, until the late nineteenth century, was done from open boats. The size of the boat dictated the number of nets that could be carried. The upper edge of the net was attached to a rope called the *back*, which was kept afloat by cork pieces threaded along its length, and, at greater intervals, usually at the joins of individual nets, by large buoys. The buoys, tarred for durability, were the inflated bladders of pigs or bullocks, or the skins of sheep or dogs. In some parts of the East Coast, dogs were bred for conversion into floats.

The lower edge of the net, known as the *skunk*, either floated free or was rigged, at intervals, with loops for attaching stone weights. The nets were shot at dusk, and the train attached to the bow of the boat by a long *swing-rope*. Boat and nets thus rode together until dawn, when hauling usually began. As the nets came in, the crew would be constantly shaking out the meshed fish.

Of the earliest Scottish fishing craft, little is known. Some would have been made of wood, some would have been skin-covered. In the eighteenth century, the boats were generally open, clinker-built (constructed of overlapping planks), low in the water, and - Viking style - pointed at both ends. In bad weather, they were entirely comfortless, the crews exposed to wind and rain and breaking seas. When caught in storms, they were also dangerous, being liable to swamp. Boats then, as now, came in all shapes and sizes, built for specific fishing methods and sea conditions. Regional characteristics became apparent in the nineteenth century, when part-decked and fully-decked boats began to evolve.

Boat-building was formerly carried on wherever there were fishing communities. Most fishing towns and villages would have several builders. Wick in 1841 had twelve yards, employing from 70 to 80 men and launching from 80 to a hundred boats annually. Sawing of the rough logs was often done by itinerant sawyers, who took on work by contract. In the treeless Shetlands boats were assembled from prepared timber imported from Norway.

The type of boat favoured on the southern shore of the Moray Firth was the *skaff* or *skaffie*, characterized by a rounded stem and a very sharply-raked stern. Elsewhere on the East Coast, the favoured type was the *fifie* (supposedly of Fife origin) in which stem

Boat-building at the yard of Smith and Hutton at Anstruther, Fife, in 1973. The carpenters are fairing, *or smoothing, the timbers with adzes to ensure that the planks to be nailed into position lie flush.* SEA

A typical Zulu, the Inverness-registered True Vine. *This model, which does not represent any actual vessel, both name and number being the maker's choice, was commissioned from Mr A Manson by the NMS.*

and stern were almost vertical. Smaller fifies, for inshore fishing, were known as *baldies*, supposedly a contraction of (Giuseppe) Garibaldi, the Italian patriot active in the mid-nineteenth century.

In 1879, still another type appeared, not by evolution, but by the calculated fusion of the upright stem of the fifie and the raked stern of the skaffie. This prototype, appropriately named the *Nonesuch*, was designed by William Campbell of Lossiemouth and rapidly displaced the skaffie from its traditional homeland. Known as the *Zulu*, after the Zulu wars, then in progress, this type has been described as 'the most noble sailing craft ever designed in the British Isles'. Both Zulu and fifie were latterly carvel-built (flush-planked) and increased in size as the century approached its end. By 1903, the largest Zulus exceeded 80ft (24.38m) and the largest fifies 70ft (21.34m). Everything about these boats was big. The hulls were massively constructed: in all, about 40 tons of undressed timber, both native and imported, went into the construction of one fifie of the largest class.

The foremast alone of a big Zulu could weigh two tons and tower 60ft (18.29m). Both fifie and Zulu carried two masts (fore and mizzen), a bowsprit and a mizzen boom, set over the stern. Sails carried were fore, mizzen and jib. As the boats grew bigger, so too did the sails, and the hoisting of the foresail was a job for all hands, excepting only the skipper. Ironically, as the splendid big luggers reached the peak of their development, they were already on the way out, overtaken by steam power.

The ring-net was an active method: fish were first located and then surrounded by the net in a two-boat operation. Though essentially a West Coast method, suited to enclosed waters, it was also used traditionally in the Firth of Forth (p46) and, later, in the Moray Firth. The earliest 'trawl' or ring-net skiffs were 20-25ft (6.10-7.62m) long, open, and usually propelled by four oars, though most carried a small lugsail. In the mid-1870s, the size of the skiffs began to increase, and in 1882 the forerunners of the true *Loch Fyne Skiff*, the *Alpha* and the *Beta*, were built for Edward McGeachy in Dalintober, Kintyre. The typical Loch Fyne Skiff ranged in length from 32 to 37ft (9.75-11.28m) and was furnished with a small forecastle, or *den*, for the crew's accommodation. The stem was curved and the stern raked, giving the short keel essential for working in the narrow waters which were the characteristic haunts of the ring-netters. The steeply-raked mast was positioned well forward to free space aft for 'boating' the net and catch.

The decline of the Loch Fyne Skiff began in 1922 with the launch of a pair of fully-decked boats, the *Falcon* and the *Frigate Bird* for Robert Robertson of Campbeltown, the pioneer of modern ring-netting. These boats, beautifully proportioned, with canoe sterns and a small wheelhouse aft, were 50ft 2ins (15.29m) long and 15ft (4.57m) of beam. Startlingly unlike the Loch Fyne Skiff, they were generally condemned as 'money lost', but by 1945 the bigger decked boats had almost completely supplanted the skiffs, and in the 1960s the model was adapted, by filling out the hull, to operate dually as trawler-ringer.

6 The cure

To cure is to preserve, and fish were preserved by salting, drying or smoking, or combinations of these processes. Canning and freezing, in modern times, have removed the necessity of preservation by salt, and salt fish is now a rarity on the table, difficult to obtain and eaten only by those, generally older folk, with a taste for it. Salt herring, in the eighteenth and nineteenth centuries, were a great staple of the poor of town and country, adding flavour to an otherwise monotonous diet of potatoes. Most families kept a barrel or two for winter provisions.

The fishermen were the first to be engaged by curers, months before the start of the fishing season. They were given a guaranteed

Curer outside curing station on the shore of Loch Carloway, Lewis, 1920s. Alasdair Alpin Macgregor, SEA

price - 11s, say, in the 1840s - for every *cran* (a measure of fresh herrings, latterly fixed at 37.5 gallons or 170.48 litres) landed up to a maximum of 200 or 250 crans. Once engaged, a crew was obliged to supply that particular curer and no other; a bounty was paid to each crew to seal the bargain. Amounts varied, according to the reputations of the skippers. One crew might be paid five guineas, another eight. Additional inducements included the promise of a weekly supply of whisky, perhaps four pints to each crew.

As a rule, three women - two to gut and one to pack - were employed for every boat engaged. These women received a small advance as *arles* (engagement money) and were paid thereafter per barrel packed. Speed was therefore the essence of their labour - the more fish they handled, the more money they earned - and in all accounts of the bustle of the curing-yards, the lightning movements of the gutters' hands are sure to be remarked on. Neil Gunn: 'I can remember as a very small boy watching a woman gutter to see if I could follow how she did it and being baffled.'

The herring catches were unloaded into *farlans*, long wooden troughs at which the gutters stood. With one stroke of the short gutting-knife, the throat of each fish was slit, and gills and gut whipped out. Not only was the gutter gutting, she was also grading each fish by sight alone. The gutted herring were mixed with salt in tubs and then packed tightly, layer upon layer, into barrels, with salt between each layer. When the barrels had been filled, the lid was fitted on and the fish left for about eight days. The barrel was then reopened and the contents, diminished by shrinkage, topped up with fish of similar size, quality and stage of cure. The barrel was again closed and sufficient brine poured through the bung-hole to fill every space. Finally, the name of the curer and the port of origin were stencilled on the bottom end. One barrel of salt was reckoned sufficient to cure three barrels of herring.

With the establishment of the Fishery Board for Scotland in 1808, it became possible for the first time to regulate strictly the

Herring gutters at Greenhill, Peterhead, about 1890.
Probably taken by James Schivas

quality of cured herring. The Scottish coast was divided into districts, in each of which at least one officer was stationed. To these officers fell the duty of sampling batches of herring. Only those herring passed by Board officials were eligible for bounty. From 1786 to 1814 the bounty was 2s per barrel. Thereafter it was increased to 4s, and in 1829 abolished.

When a batch of cured herring had been inspected and passed, it received the Crown Brand: a crown, the word 'Scotland', and the category of the selection. The Crown Brand was regarded abroad as a sure mark of quality. There were seven categories of cured fish, defined according to size and condition. A barrel of *matties*, for example, consisted of young maturing fish of not less than nine inches (0.23m) long and with the long gut removed.

James More's herring-curing yard at Shaltigoe, Wick, 1935.

Not only were fish inspected, but the barrels themselves, which had to meet meticulous specifications as to size and construction.

Wherever a major herring fishery opened, the curers were sure to follow. By 1884, within four years of the great expansion of the Shetlands fishing, 112 curing stations had been erected. These stations were quickly organized: a jetty constructed and prefabricated huts assembled for the labour force, close to the open yard in which the work was carried on. One such station was described by E W Hardy in his *Life and Customs in the Shetland Islands*, published about 1910:

> The fishing stations are not pretentious places, and a casual glance shows only a confused mass of barrels, tanks and sheds. The sheds are used as temporary homes for the girls who clean or 'gut' the

herring. Each room ... is occupied as a rule by six persons. The fish curer provides a table and sufficient seating ...a handy little stove, and fuel. There are but three beds, or sometimes four, but they are most commodious, and amply big enough to accommodate more than the two girls who share each bed. When the room is neatly papered, and a few picture post-cards and almanacs hung on the wall, it presents quite an attractive appearance...

Girls were drawn from all parts of Scotland and beyond to the seasonal stations, and where the fleets moved, they followed, all the way south to the great autumn climax of the herring harvest, East Anglia. Hundreds of Scots lassies, each with her wooden chest, or *kist*, arrived in Yarmouth and Lowestoft by special trains. In the streets, they were unmistakable, clad in brightly-coloured woollen jumpers, thick skirts, leather 'half-wellingtons', oilskin aprons and coloured shawls, and when they weren't gutting they'd be knitting.

These teams of girls often held together for many seasons, until marriage broke up the partnership. Some left home at the age of fourteen to become gutters. To protect themselves from cuts, the gutters bound their fingers with cloth strips. But despite these precautions, accidents did happen, and should a wound begin to fester, a poultice of bread and sugar was the remedy. The girls were also expected to assist the coopers in the strenuous task of stacking the sealed barrels, which labour often revealed its damage during later child-bearing.

A big proportion of gutters and packers came from the Western Isles, particularly Lewis. In 1907, more than 2,000 Lewis women worked at the East Coast and Shetland fishings, and more than 1,000 of the same women at the later English fishings. Their combined earnings amounted to £33,200, a vital contribution to the crofting economy of the island.

The coopering trade was essential to the curing industry. By the late nineteenth century, Aberdeen, Fraserburgh, Peterhead and Wick were its centres. The vast bulk of the wood was spruce from Norway and Sweden, but small amounts of birch, ash, elm and alder were used. Out of season, coopers were paid per barrel made, but during curing operations they received a weekly wage. As the curing industry approached its record year, 1907, and the demand for barrels increased - in 1906, Scottish coopers turned out no fewer than 2,094,014 barrels and 422,080 halfbarrels - mechanization began to take over. Seven factories were operational by 1907: three in Fraserburgh, two in Peterhead, one in Aberdeen and one in Lerwick.

Where were these herring destined, preserved in death in their tens of millions? In the eighteenth century, the main export market was the West Indies, but with the emancipation of slaves in 1833, the 'food of negroes' lost its hold there. Exports outside Europe, primarily to the West Indies, entered a steep decline between 1831 (57,128 barrels) and 1837 (34,142 barrels) until by 1847 merely 3,700 were shipped. Ireland was another important

Barra herring-gutters at Ullapool, 1929. SEA

market, dented by the famine years of the 1840s. Between 1831 and 1845, the number of barrels of gutted cured herring shipped there only once fell below 100,000, and in 1841 reached 186,747 barrels. In 1846, the number fell to 98,346 barrels.

The European markets, conversely, expanded fairly steadily between 1831 (31,100 barrels) and 1841 (91,068) and in 1842 broke the 100,000 barrier. By the mid-nineteenth century, the main countries supplied were Germany, Russia and Poland. These inroads, forced upon the curers by the collapse of the West Indian trade, were at the expense of the Dutch and Norwegians. Thereafter, the Continental trade boomed, annual exports increasing from an average of 224,655 barrels in 1851-1855 to 1,143,207 in 1881-1885. But in 1884 a crisis of over-production brought on recession. Curers were forced out of business, hundreds of boats laid up, and coopers laid off. Recovery took almost

Coopers, with the tools of their trade, posed in the Johnston Studio, Wick, about 1865.

ten years, and was followed by another boom, which climaxed in 1907 and ended with the outbreak of war in 1914. The curing trade never wholly regained its European markets after that catastrophic disruption, and in the end both Germany and the Soviet Union intensified their own fishing efforts. In 1970, merely 19,900 barrels of herring were cured in Scotland, and 186 persons and sixteen gutting machines employed at eleven stations. The main buyers were East Germany and South Africa.

The kipper is a smoked herring, but the smoking process is more for the sake of flavour than of preservation. The fish is split open, cleaned, immersed in brine for about an hour, then hung overnight in a smoke-house fuelled by oak-chips and sawdust. The production of 'red herring', by dry-salting and oak-smoking, could extend to three months.

7 Salmon

Historically, the salmon must rank as the most important of fish, more important even than the herring, for the nature of its life made it so. Stated simply, salmon that are born in Scottish rivers migrate into the ocean and return to these same rivers, themselves to spawn. From earliest times, people would have watched for the returning fish and taken what they needed for food. The supply was dependable, conveyed itself deep into the countryside, and one large fish would feed a family.

Pollution, a major by-product of the Industrial Revolution in the eighteenth century, and over-fishing in the nineteenth century together spoiled many Scottish rivers for salmon. But before that happened, and salmon became a luxury food, there was generally abundance for everyone. There are many reports of the immense riches of the Scottish waters. In the 1650s, Richard Franck remarked, of the Forth at Stirling, that 'the abundance of salmon hereabouts is hardly to be credited'. From the Sluie Pool on the River Findhorn, in June 1648, no fewer than 1,300 were taken in one night, and Thomas Pennant records, at

Thurso in the 1770s, 'not less than 2,500 taken at one tide within the memory of man'.

As a source of revenue as well as of food, the salmon fisheries were given protection - on paper, anyway - in a succession of laws going back to the eleventh century. King Robert III, in 1400, introduced capital punishment for anyone thrice convicted of killing salmon during their spawning season, but whether that extreme penalty was ever carried out remains doubtful.

Along with wool and hides, salted salmon was a major Scottish export during the Middle Ages. Aberdeen, Perth, Glasgow and Berwick were already centres of the salmon trade by the thirteenth century, and French, German and Italian merchants journeyed to Scotland, bringing cloth, silk, velvet, spices and wine, and taking away salmon. Salmon were being shipped to Flanders and France at least as early as 1380. A large market for Scottish salmon also existed in England, but was subject to periodic disruption when these two inimical peoples were quarrelling or at war.

'Piece'-time in a salmon-fishers' bothy on the North Esk, about 1914. SEA

In 1786, a Scottish fish-merchant, George Dempster, began shipping salmon in ice from the River Tay to London. The innovation caught on rapidly, and by the end of the century ice-houses had been built at the main salmon-fishing centres of Scotland. Until then, all fish for export were cured with salt in barrels or boiled and then pickled in *kits* (small barrels) with vinegar as the preservative. The big ice-house at Tugnet, Speymouth, was built in 1830 and continued in use until 1968. Thickened ice, for storage, was broken from the river and from special ponds in winter. The use of ice did not, however, immediately eliminate the older forms of preservation. Come June, and warmer weather, icing was reduced and pickling increased, and when the London market had been 'plentifully supplied' with the pickled product, salting for foreign markets would begin. Smacks were kept running from the Tay to London carrying iced fish and returning with 'porter, cheese, groceries and other goods'. The trade in iced salmon became so all-consuming that, by the end of the eighteenth century, there were complaints from many places that natives were unable to buy the fish locally.

The methods of catching salmon were, and are still, many and varied and the following is just a selection of these. Seine-nets, set by boat in a circle and hauled from the shore, were in use as far back as the Middle Ages. It was remarked of the Speymouth seine-fishermen in the late eighteenth

Salmon leisters.

A net being hauled on to drying poles from a punt at Kames in the Kyles of Bute, nineteenth-century. Detail from a painting by David Fulton

Net-mending at Newhaven in the nineteenth century. The girl holds a creel containing herring, and cod have been arranged on the table. Detail from a painting by Alfred H Green.

Hauling a seine-net to the shore at Kirkcud-bright, late nineteenth century. Detail from a painting by W S Mac-George.

'Burning the water' at a salmon pool. The man on the left is poised to strike a fish with his leister, while his companion holds the torch and lugs the creel into which the speared fish are flung.

The profusion of gear on board the Carradale clam-dredgers
Comet *and* Monarch, *in Campbeltown harbour, 1994.*
The dredges can be seen laid along the port sides, in readi-
ness for shooting.

The Altaire *of North-mavine, Shetland, newly launched from the yard of Langsten Slip & Botbyggeri AS, Tomreford, Norway, 1994.*

Unloading the catch at Eyemouth, 1989.

century that, 'the hardships which these people undergo ... in wading in the water, often above the knee, during the winter and spring, and remaining in wet clothes perhaps for 12 hours, will appear almost incredible to those who have not witnessed it'. The netting of rivers and estuaries was organized on capitalistic lines. The landowner let his fishings to *tacksmen* (lessees), who then supplied the *cobles* (boats) and nets, and hired crews. Until the late eighteenth century, the *currach* (coracle), an oval, wicker frame with a cow's hide stretched over it, was in use on the Spey for setting nets.

Cruives were fish-traps of wood or wicker, built into weirs and positioned to intercept the progress of fish upriver. Ancient cruive-dykes can still be seen on several rivers, for example on the Forth above Stirling.

Fish-spears, called *leisters*, were commonly used. At Rathven, near Forfar, in the eighteenth century, 'black-fishers' speared torpid spawners in the gravelly shallows by night, 'preceded by a great torch, or blaze as it is called, consisting of dried broom or fir tops, fastened round a pole ... The strongest constitutions often suffer from wading up to the middle in water for hours together ... and a black-fishing match often ends in drunkenness and debauchery.'

Walter Scott has described salmon-fishing in several of his novels, notably *Guy Mannering*, in which the hunters, 'like the ancient Bacchanals in their gambols, ran along the banks, brandishing their torches and spears'. Upwards of a hundred fish were divided among the participants, the best going to the 'principal farmers' and the rest to the 'shepherds, cottars, dependents, and others of inferior rank', who dried their share 'in the turf smoke of their cabins or shealings' as a 'savoury addition to the mess of potatoes, mixed with onions, which was the principal part of their winter food'.

The dip-net, a bag suspended from a circular frame and attached to a long pole, was in general use, while on the River Nith, in the Solway Firth, the traditional *haaf-net* is favoured.

Haaf-net fishers at work in the River Nith. SEA

This is a large portable bag-net, lashed to a rectangular frame, operated by a fisherman who stands chest-deep in the tidal channels and waits for a fish to enter the net.

Both *stake-nets* and *bag-nets*, the former secured on the seabed by stakes and the latter floating, are fish-traps operating on the same principle as the old cruives. Both are used to intercept salmon on the coast, and became popular in the first half of the nineteenth century. Driftnet fishing for salmon in the open sea became illegal in Scottish waters in 1962, after a brief existence.

Much modern poaching is inexcusable in its commercial ruthlessness: dynamite and cyanide have been used to kill large numbers of fish in one swoop on a river. But the average poacher's expectations are modest - 'one for the pot' - and his tools simple: a wire snare or *cleek* (gaff with which to hook the fish by its belly or back). Cleeks are sometimes completed at the riverside by cutting a shaft and lashing it to the hook. Twine and the hook are therefore all that the poacher needs to conceal in his pocket.

Angling for salmon, though not strictly a commercial fishing method, is nevertheless of great economic importance to the tourist industry in Scotland, and gives pleasure to tens of thousands of natives, when allowed the privilege of casting on the waters. Some remarkable feats of rod-fishing have been recorded. The record for catching prowess apparently belongs to one Alfred Denison who, from 1860 to 1887, fishing the Tweed and the Ness, landed a total of 4,600 salmon and *grilse* (a salmon which spawns in the year after leaving the river).

Salmon-farming in Scotland began in the late 1960s and expanded rapidly in the 1980s. It is now a significant industry on the western coasts and islands, where suitably sheltered sites for the mooring of rearing cages are found. The industry has not, however, lacked its critics, particularly in the matter of environmental impact through seabed pollution by excess feed, chemical wastes and medications. Another concern is the genetic contamination of wild stocks by fish-farm escapees. Some observers believe that the negative impact of salmon-farming outweighs the economic benefits it generates, while others maintain that salmon-farming satisfies a market which wild stocks can no longer supply, and therefore relieves pressure on these stocks. In 1991 salmon-farmers in Scotland harvested 40,600 metric tons of fish. In that same year, 228.1 tons were landed by rod and line, 97.2 tons by net and boat, and 137 tons by 'fixed engine' (stake- and bag-nets).

8 Garvie, mackerel, spirling, powan

The sprat is a close relative of the herring - indeed, young herring and sprats often swim together - but it is the smaller of the two, rarely exceeding six-and-a-half inches (0.16m) in length, and is more of an inshore fish. Its Scots name is *garvie*. The main Scottish sprat fishery was in the Firth of Forth. Drift-nets were used in the shallow estuary west of Queensferry, and at times as far up the Firth as Kincardine and Alloa, where another, static method was used. The cruive was a long, funnel-like wicker trap. Its mouth faced upriver, to capture sprats carried down on the ebb tide. In the late eighteenth century there were 83 cruives at Longannet and 61 at Kincardine. The fishing season lasted from August to early March.

The net fishery usually began in November and lasted until the end of February, a short season, but economically vital to the fishermen of Queensferry and Newhaven. The garvies, 'a cheap food, much relished by the poor', were sold locally and exported to England. Drifts were in use at first, but 'trawl-nets' - actually beach-seines - appear to have been adopted in the late 1820s. In 1851, the Forth garvie 'trawls' were outlawed along with the Loch Fyne herring 'trawls' (p24). As in Loch Fyne, there were violent confrontations between angry fishermen and the crews of fishery cutters and Royal Navy ships sent to police the Forth. In November 1860, for instance, a boat's crew from HMS *Lizard* was stoned and shot at from Queensferry harbour. Nets and boats were seized and sold by public auction, and fishermen were fined. Until 1968, most Scottish sprats were caught in the Firths of Forth, Tay and Moray, but in that year a West Coast fishery with mid-water trawls was opened up.

The mackerel is one of those species, like conger eel and dogfish, which are esteemed in some communities and despised in others. It is, like the herring, a pelagic (surface-living) fish, and likewise swims in shoals. Mackerel-fishing in Scottish waters was slow to develop, owing to the difficulty of securing markets. In the 1890s and early 1900s, a lucrative trade in cured mackerel for the American market developed, particularly in the Clyde and Minches. Drift-netting was the main method, but hand-lines were also used. From the 1930s on, the centre of mackerel-fishing in Scotland was Fraserburgh district, where up to 100 boats were engaged annually in summer, working *darras*. These hand-lines with feather lures were also known as *jigs*, from the jigging, or jerking, technique used.

Mackerel-fishing on a hitherto unprecedented scale, primarily with purse-seines (p79), began in the late 1970s. The Scottish catch increased from 17,000 tons in 1975 to 54,000 tons in 1977, and continued to increase. The immense migrating shoals are netted west of the Hebrides and transshipped, chiefly at Ullapool, on Loch Broom, to large Eastern European vessels, known as *klondykers*. The system of klondyking began in the early 1900s with the shipment of lightly-salted or iced herring to the Continent, chiefly Germany.

The *spirling* or *sparling* (smelt), a shoaling fish, was caught only in the Forth and Tay estuaries and in the Solway Firth, mainly in bag-nets. Described curiously in the eighteenth century as tasting and smelling strongly of rushes, sparlings were considered a delicacy and commanded a ready market, selling in 1895 at 67s 9d per hundredweight compared with 4s 6d per hundredweight for herring. It has to be said, however, that supplies were never abundant on the market. The Tay fishermen, 90 of them in 1893, generally sent their catches (297 hundredweights in that year) to market in small boxes containing three pounds of fish. In the early twentieth century, landings seldom exceeded 200 hundredweights annually.

The *powan*, sometimes known as the 'freshwater herring', is, in Scotland, found only in Loch Lomond and Loch Eck in Argyll. The *Old Statistical Account* for Buchanan on Loch Lomondside records that in the late eighteenth century powans were 'at times caught in great quantities on this side of the loch with draught (i.e. seine) nets'. The powan fisheries were revived during the food shortages of the Second World War.

9 Shellfish

The only species of oyster native to Britain is the common oyster, which is essentially a shallow-water species. Oysters have been gathered as food since earliest times, but in the eighteenth and nineteenth centuries important fisheries developed on natural oyster beds. Until the nineteenth century, oysters were a food of the poor. Thereafter, as demand exceeded supply, they became, as now, a luxury food. They were hawked with the cry *Caller oo!* (fresh oysters) in the streets of Edinburgh and sold in the 'oyster-cellars' of the city, celebrated in the late eighteenth century by Robert Fergusson:

> ... And sit fu' snug
> oe'r oysters and a dram o' gin
> or haddock lug.

The main Scottish oyster fisheries were located in the Firth of Forth and in Loch Ryan, but smaller fisheries developed at West Loch Tarbert, Wigtown Bay, Lewis and the Orkneys and Shetlands.

In 1773, when a Leith merchant was contracted to ship oysters to London, sixteen dredgers were operating from Cockenzie, sixteen from Prestonpans, Cuttle and West Pans, and eight from Fisherrow. The oysters were packed in barrels, and twelve vessels, each with an average cargo of 320 barrels, were employed from mid-January until mid-May in the trade. The Glasgow market was also supplied, by horse and cart.

At that time, the *scalps* (beds) were so productive that 6,000 and more oysters could be dredged daily by a single boat. The Forth beds, however, became 'greatly exhausted' by the late eighteenth century, and the fleet of dredgers declined to ten boats, and catches to 400 or 500 oysters per boat. Profits were divided by six, one share for each of the five-man crew, with the extra share going towards the boat's upkeep.

Rowing songs, to aid the rhythm of the oarsmen when there was insufficient wind for sail, were sung on the Forth oyster boats. Though these 'dreg sangs' were essentially improvised, some, like the following extract, became standardized.

> The oysters are a gentle kin,
> They winna tak unless you sing.
> Come buy my oysters aff the bing   off the heap
> To serve the shirreif and the king,
> And the commons o' the land
> And the commons o' the sea,
> Hey benedicite! and that's good Latin.

Overfishing of the oyster beds led to serious decline, and the last traditional fishery in Scotland was that in Loch Ryan. In the present time, however, oysters are farmed along the West Coast.

The only shrimp fishery of any significance in Scottish waters was in the Solway Firth. The shrimps fished there were of the brown variety, *Crangon crangon*, and were caught in small-meshed beam-trawls, towed by small cutter-rigged smacks, each crewed by two men. In 1893 a fleet of 51 of these boats operated by utilizing the tides. The boats were cast off at the Waterfoot, Annan, with the first of the ebb, the nets were shot, and then the boats drove to seaward with the tide for eight or ten miles. The flood tide brought the fleet back over the same ground. The shrimps were boiled on board, with salt added, to be hand-peeled later in cottage 'pickeries'. In 1994, there were ten motor-powered beamers operating out of Annan. In the 1960s and 70s, North-East boats engaged in sporadic fisheries for pink shrimps (*Pandalus borealis*) on the Fladen Bank, between Scotland and Norway.

Low down on the shore, among the rocks, crabs and lobsters can be caught. Lobsters were traditionally captured by wading into the sea at low tide and extracting them from their crevices with a bent stick, rather like a shepherd's crook, or a sickle. Baited hoop-nets, similar to those described on page 9 for saithe-fishing, were also used.

Commercial lobster-fishing arrived in Scotland around the mid-eighteenth century, stimulated by English merchants. The fishermen usually entered into agreement with these companies, in some cases for up to five years, and were provided with small skiffs and gear, to be paid for by instalments. The creels used in the fishery - 20 and more per boat - were a wonder. At Duffus in Morayshire, in the 1790s, it was remarked that 'no lobster traps were ever before seen on this coast'. They were set and lifted hourly, more or less, from sunset to sunrise. Nowadays, lobster-fishermen generally lift their creels - which might, however, number hundreds - only once in 24 hours, after dawn.

Catches were, as now, generally contained in floating chests, or 'keep-boxes', and collected, usually weekly, by welled smacks and carried to the London markets. These smacks contained a chamber in which seawater could circulate, through holes in the hull, thus preserving at least a part of the catch alive.

By the 1790s, no fewer than 60 boats were creel-fishing from the Orkneys and catching annually from 100,000 to 120,000 lobsters. The landings in these early years, taken nationally, were astronomical, because the fishing grounds were virgin. But overfishing soon took effect, and by the early nineteenth century some coasts were virtually unproductive. At Fordyce in Banffshire, for instance, by 1843, only a 'few' lobsters were being caught by 'aged' fishermen and sold locally. By then, however, the West Coast had been opened up.

The lobster-creel is a trap. Lobsters are scavangers, and fish-bait lures them into the trap. They enter the creel through the *eye*, a netted passageway which is easier of access than of exit. Creels used to be made by the fishermen themselves, during winter, when they were not fishing, but increasingly are manufactured commercially in plastic. Creels can be set in fleets of 20 or so, strung together on a single main rope, for ease of recovery.

Lobster-fishing continues to be worked mainly by part-time fishermen in small boats, and fits well into the routine of crofting. Mechanical pot-haulers have removed the labour of pulling in creels by hand, and many modern fishermen operate fast boats, thus increasing their range, a decided advantage now that the nearer shores yield few lobsters.

Historically, the only species of crab eaten in Scotland was the large edible crab, which has a reddish shell. A specimen, weighing six pounds and measuring eighteen inches (0.46m) across the back, was caught by Crail, Fife, fishermen in 1994 and presented to St Andrews Sea Life Centre for live exhibition. In recent years, smaller species of crab have become marketable on the Continent. The fishing of edible crabs has always occupied East Coast fishermen more than West, owing to the proximity of markets. In 1908,

94 per cent of the total landings - 2,736,712 crabs - were caught on the East Coast, chiefly from Montrose district south. By the 1960s, the bulk of landings were from the Berwickshire coast, Firth of Forth, and around Peterhead, Fraserburgh and Wick.

There are four species of scallop in Scottish waters, of which two are eaten. The biggest is known in Scots as the *clam*. The lower, concave, shell used to serve as a milk-skimmer in butter-making and as a receptacle in whisky-drinking - indeed, there is a *piobaireachd* (classical bagpipe) tune titled '*Bodaich Dubha nan Sligean*' (the Black Old Men of the Shells). Much further back, in mesolithic times, clam-shells were in domestic use, probably as scoops and containers. Nowadays, a clam-shell is more likely to be seen in use as an ash-tray. The clam is now a delicacy, and one of the most expensive of shellfish; yet, its commercial exploitation in Scotland began only in 1936, in the Clyde.

The main method of clam-fishing is by dredging, though quantities are collected commercially by divers. The clam-dredge

Lobster-fisherman Donald Wares, baiting a creel off Dunnet Head, Caithness, 1960. SEA

is a triangular iron frame with a toothed crossbar or *sword* set at an acute angle for raking the shells out of the seabed. Clam-boats initially towed from two to six dredges, but with increasing engine-power and decreasing stocks, seven dredges are now generally towed per side. The development of the fishery accelerated in the mid-1960s with the establishment of factories for removing and packing the meats, which could by then be flown - fresh or frozen - to Continental markets, chiefly in France. Hitherto, clams were sent to Billingsgate in the shell.

In the late 1960s, a fishery opened up for the queen scallop. The virgin inshore beds were astonishingly prolific: the research ship *Mara* once caught 10,000 queens in one dredge after two minutes' towing, in the Clyde. But there had been earlier fisheries. In the nineteenth century, queens were being dredged, mainly in the Firth of Forth, for line-bait, but that fishery declined with the decline of line-fishing. Queens can be distinguished from clams by their bright colours, daintiness, and the fact that both shells are concave. They are fished using dredges and trawls sturdily rigged to minimize damage on rocky seabeds.

The now-important trawl fishery for Norway lobsters - colloquially, 'prawns' - is essentially a post-war development. Previously, the creatures were regarded as a nuisance when taken with white-fish. The fishery's expansion was rapid: landings jumped from 2,994 hundredweights, valued at £5,539, in 1950, to 192,493 hundredweights, valued at £5,010,564 in 1973. The main trawling areas, in the early years, were the Firth of Forth, Moray Firth, Minches and Clyde, but creels were - and are still - used, particularly in the Minches, for taking prawns on rough grounds, unsuited to trawling, and in areas closed to trawling.

The prawn is a hard-shelled, pink-coloured crustacean, with slender claws. Its habit of burrowing in mud has been instrumental in preserving stocks from serious overfishing, for in strong tides and bright sunlight it proves elusive. Catches are generally handled in factories, becoming scampi in the process, but good-sized prawns, like lobsters, may be marketed entire.

10 Whaling and sealing

Stranded whales, remains of which have been found in the Forth Valley, particularly near Dunmore and Stirling, supplied food, oil and 'whalebone' as far back as mesolithic times. The harpooning of whales in Scottish waters was rare, until the present century, but whales were hunted, and hunted with great passion and determination, particularly in the Northern Isles. Such hunting was opportunistic: if whales appeared offshore, the people would pursue them and drive them aground. Whales were sometimes secured effortlessly - they simply ran themselves ashore. Martin Martin, in the late seventeenth century, reported that about 160 'little whales' grounded themselves on Tiree, 'very seasonably, in time of scarcity, for the natives did eat them all and told me that the sea-pork is both wholesome and very nourishing meat'. At Northmaven, in the Shetlands, the islanders normally eschewed whale-meat from 'a false refinement, or sense of delicacy', but in 1741, 'by reason of the extreme scarcity then prevailing, some families were induced privately to make use of their flesh, and all such declared it to be equal to any other beef'.

From time to time huge schools of whales were driven ashore: in the Bay of Hillswick, Shetland, 360 whales were taken in 1741, and on the west side of Eday, Orkney, 287 in 1841. The oil from that latter school realised £398, a huge sum then. These whales were no doubt pilot whales, popularly called *caain* whales (Scots *ca*, to drive, from the custom of driving them ashore by beating oars, throwing stones, shouting, etc.). They can reach 28ft (8.53m) in length and are still slaughtered, more for sport than from necessity, by the Faroese.

A description of a typical whale-hunt survives from the pen of Robert Scarth, of the parish of Cross and Burness in the Orkneys, in the 1840s:

All boats are launched, all hands active, every tool which can be converted into a weapon ... from the roasting spit of the principal tenant to the ware (seaweed) fork of the cottar, is put into requisition. The shoal is surrounded, driven like a flock of timid sheep to shallow water on a sandy shore, and then the attack is made in earnest. The boats push in, stabbing and wounding in all directions. The tails of the wounded fish lash the sea, which is dyed red with their blood, sometimes dashing a boat to pieces. The whales in dying emit shrill and plaintive cries, accompanied with loud snorting...

Industrial whaling in Scottish waters began in 1903. Five companies were involved, and Norwegian influence and investment were strong. There were five whaling stations - four in Shetland and one in Harris - all of which were closed with the outbreak of the 1914 war. During that thirteen-year period, 6,934 whales were killed. When, in 1920, the whaling ban was lifted, two of the Shetland stations - Olna and Colla Firths - and the Harris station, Bunaveneader, reopened. In 1929, the last of these stations, Olna Firth, closed. Total whales killed: 3,260.

The Harris catchers hunted mainly around St Kilda, while the Shetland catchers worked in the waters north and north-west of the islands, an area which, in 1923, was computed to be the most productive in the entire North Atlantic. Most of the shore workers were locals, but the hunters themselves were Norwegian. The main product was whale-oil, but fertilizer, bone-meal, dried and salted meat, whalebone and cattle and poultry food were also produced. The main species taken was the finner whale, but sei, blue, northcaper, bottlenose, humpback and killer were also harpooned. Bunaveneader reopened in 1950 and operated for two seasons with one catcher, crewed largely by Norwegians. In June 1952, having killed 53 whales, the company went into liquidation, thus bringing to an end shore-based whaling from Scotland.

Distant-water whaling from Scotland began in the mid-eighteenth century, stimulated by government bounties on whale-oil. The first subsidized whaling trip was evidently from Leith, in 1750. In 1751, two whalers fitted out at Campbeltown, and in 1752 Dunbar and Greenock entered the business. Aberdeen, Dundee, Bo'ness, Kirkcaldy, Peterhead, Port Glasgow, Montrose, Fraserburgh, Banff and Burntisland followed. Aberdeen,

Dundee and Peterhead were the main Scottish whaling ports. Indeed, by the mid-nineteenth century, Peterhead was the main whaling port in Britain.

The method of hunting was with hand-held harpoons from small, open rowing-boats, no job for the faint-hearted. A harpooned whale was played, usually by several crews, on miles of rope until exhausted, then finished off with lances, a hideous climax which could take many hours. Whaling was chancy, yet few whales were needed to ensure the success of a trip. The *Success* of Dundee, for instance, took eleven whales in 1791, and that was one of the best catches of the season among the Scottish whalers.

Whales were flensed - stripped of their blubber - alongside the ship, and the blubber cut into strips and packed in barrels. The malodorous oil-extraction process, by boiling in copper vats, was done when the ship returned to port at the season's end. Whale-oil, the main product of the industry, was largely used for lighting and lubrication. Baleen, the plankton-straining plates, often

Flensing a blue whale on the slip at Loch Bunaveneader, Harris, May 1952. Angus M MacDonald, SEA

erroneously termed 'whalebone', was a valuable by-product, mostly used for stiffening corsets and hooped dresses.

The main prey was then the bowhead whale, also called the Greenland right whale, because it was the 'right' whale to kill, being blubber-rich, buoyant in death, and, in its slowness, relatively easy to kill. As one ground was hunted to depletion, another ground took its place, the whalers penetrating further and further into the bone-chilling ice-fields of the Arctic, in which their ships were frequently trapped and crushed. Four Aberdeen whalers were lost in 1830, a catastrophic year in which nineteen of the 91 British whalers in the Davis Strait were destroyed in the ice, and 21 returned 'clean', or empty.

Conditions were arduous in the extreme. Whether rowing for hours in pursuit of whales, slaughtering seals on unstable icefloes, or ice-breaking to make a passage for their ship, the men were constantly wet and cold. Wages comprised a monthly fixed sum with a bonus related to the quantity of oil produced. In a bad season, some crews might finish a trip in debt, after deductions for coal, tobacco, soap and clothing, and find themselves compelled to sign on for another trip in order to pay back the money. Their basic diet consisted of salted beef and pork, bread, flour, oatmeal and other cereals, butter, cheese, spirits and beer. Tea, coffee and sugar were supplied on some ships, but on others the crews had to bring their own.

The most distinguished Scot to have sailed on an Arctic whaler was no doubt Arthur Conan Doyle, creator of 'Sherlock Holmes'. He signed on as ship's surgeon in the *Hope* of Peterhead in 1880, while still a student at the University of Edinburgh, and spent seven months at sea. He was evidently so highly thought of by the captain that he was offered double pay, as harpooner as well as surgeon, if he would sign on for another trip; but he declined the offer. His experiences in the Arctic made a lasting impression on him, but he considered the work brutal.

Dundee was the last major Scottish whaling port. The industry there experienced a revival in the early 1860s, with the expansion

The whaler Eclipse *of Peterhead in the Arctic, with a capture alongside. She was built in Aberdeen in 1867. Like all whalers, she was of remarkably strong construction, to enable her to penetrate pack-ice. She carried eight whale boats and a crew of 56 men, many of whom were shipped at Lerwick on the outward passage. The NMS has a scale model of the ship, made by Mr J Gray, son of her master, Captain David Gray.*

The Dundee whaling-master, Captain William Adams, in the crow's-nest of the Maud, *1889.*

of the jute industry. The price of whale-oil, used in the softening of jute fibre, rose from £30 to £50. From eight whalers in 1861, the fleet doubled by the mid-1880s. The ships were then steam-powered and faster, and a bow-mounted gun, which fired an explosive harpoon, had been invented. That innovation ultimately eliminated handharpooning from small boats, and enabled the fast and powerful rorqual whales to be pursued successfully for the first time. Right whales had by then been hunted almost to extinction in the North Atlantic.

But the whaling industry did not die then. It shifted its operations to the Antarctic, following the initiative of a Norwegian, Captain C A Larsen, who took the *Jason* south in 1892, to prospect for seals and right whales. Four Dundee whalers, led by the *Balaena*, prospected there in the same year, and the Scots and Norwegians actually met in the Weddell Sea. The Scots returned the following summer with seal-oil and skins, but no whales, and the trial was not repeated.

Larsen, however, persisted, and in 1904 established a whaling station in South Georgia. The Scots in due course followed. In 1908, Christian Salvesen and Sons of Leith began operations from South Georgia and remained there until Antarctic whaling was halted in 1963. The waters around Antarctica are now a sanctuary for whales, and whaling itself is a major ecological and moral issue, which will not be resolved until all whaling has ended.

Whales were not the sole prey of whalers. From the earliest years, seals were killed on the Arctic ice-floes, often in vast numbers. In 1822, for instance, the *Union* of Peterhead caught five whales and 2,500 seals, and in 1847, two whales and 7,500 seals. Seal-killing was a brutal business, carried out with guns, but more often with clubs. Seals too yielded oil, which was only slightly less valuable than whale-oil, and the skins for a time were in demand for clothing and leather. Seals were also hunted on the coasts of Scotland. There are two native species, the common and the grey, which latter traditionally bore the brunt of the hunters' clubs, owing to its habit of congregating on breeding shores and islands. Seals were hunted all around the Scottish coast, but particularly in the Western and Northern Isles.

In the eighteenth century, up to 50 seals were clubbed to death annually, around Michaelmas, in a sea-cave at Gress in Lewis. The cave was entered by boat, and the approach completed on foot, the men carrying torches lit from a pot of live coals. In some places, for example North Ronaldsay in Orkney, wide-meshed nets were used to capture seals. By the early nineteenth century, seals were also being shot. Sunken corpses were retrieved, in the Shetlands, by dogs or by use of the *klam*, a rope-operated grab. The flesh of both species of seal was eaten by poor folk, the 'young and tender ones' being preferred.

Fishermen, and in particular salmon-fishermen, have never enjoyed friendly relations with seals, owing to their predations on netted fish and to the damage they do to nets. Some salmon-fishing companies paid out bounties on all seal-tails delivered. The controlled culling of seals was discontinued in 1985, but in recent years fishermen in general have expressed alarm at the increasing populations of grey seals and have been pressing for a resumption of culling. But the issue is a highly emotive one, which, if pressed vigorously, will bring the industry into conflict with increasingly powerful conservation bodies.

11 Weather lore and wreckings

Before steam and motor power arrived, fishermen were dependent on sail and oar, and there was far less margin for error in their judgements; but weather signs were there to be read by those who could read them. Sometimes the signs would be misread or perceived too late, with fatal consequences. The signs were in the sky and in the water itself, and the crying of certain birds gave warning of hard weather. For example, the call of the great northern diver foretold a gale of southerly wind, and fishermen would haul their gear and run for shelter.

The more familiar signs in the sky were heeded, particularly cloud formations and the proverbial red sky in the morning, along with less common phenomena, such as the sun-dog, which was the stump of a rainbow, appearing with a distant shower of rain and judged to be a portent of broken weather or gales.

The appearance of a *bound* (a heavy rolling motion on the sea) on a windless day often signified a gale approaching from that direction; and, again citing an example from the minutiae of weather lore, ripples, or 'wee squalls', within the troughs of waves when a breeze was 'on', were interpreted as indicators of an imminent southerly gale.

The Shetland haaf fishermen were able to return to land from great distances in the densest of mist by following the *moderdai* (mother wave), 'a surge which drives landward no matter what direction the wind blew', as the Shetland fisheries historian C A Goodlad put it. The ability to 'read' that sign has apparently since been lost.

Not a year goes by, even now, without the loss of fishing boats and their crews. Modern fishermen will work in weather conditions which their forefathers would not have put to sea in. The boats nowadays, and especially the bigger boats, are better designed and equipped for hard weather operations, but there is an economic factor involved. A modern fishing-boat which is not at sea is not earning, and a boat which is not earning is accumulating a lot of debt. In the past, a boat could be tied up for weeks at a stretch in bad weather or periods of slack fishing, and the only expense involved would be the provisions which the crews needed to keep themselves and their families alive; and these

The conditions in which present-day fishermen at times have to work. In storm-force wind and 50ft swells, the Lynnmore *of Burghead wallows 30 miles east of Orkney, taking in water. This dramatic photograph was taken on 19 September 1990 from a Coastguard search and rescue helicopter which conveyed a pump to the stricken vessel and remained with her until her engine-room had been pumped out.* Kieran Murray

would usually be supplied on credit until the boats began earning again.

There is more than weather to threaten fishing-boats nowadays. NATO submarines on excercise in the Clyde area alone have been involved, since 1974, in some fourteen known incidents with trawlers. One of these incidents proved fatal to Jamie Russell and the three-man crew of the *Antares* of Carradale. She was snagged by her trawl-warps off the north-east coast of Arran on 22 November 1990 and dragged under. The bodies of her crew were all subsequently recovered by the Royal Navy, as was the vessel herself. She is now on display at the Scottish Maritime Museum, Irvine.

The loss of the *Antares* was the first fishing tragedy to afflict the village of Carradale since March 1911, when the skiff *Mhairi* capsized and sank in darkness during a sudden squall off the Kintyre coast. The entire crew of four belonged to the McIntosh family - Walter McIntosh, with his two sons and a nephew - and their bodies were never recovered. Such a loss, though unquestionably devastating to the family involved and to the wider community, could have borne no comparison to the catastrophies which periodically visited the fishing communities of the Shetlands and East Coast. That no major loss of life occurred on the West Coast - as far as records inform us, anyway - is entirely attributable to the sheltered nature of that coast. But, once round Cape Wrath, there is open sea and exposed coast, and such harbours as existed in the nineteenth century and earlier were mostly inadequate and difficult of access.

The Shetland fishing communities suffered two appalling disasters in the nineteenth century. On 16 July 1832, seventeen sixerns and 105 men were 'overtaken by a dreadful storm and buried in the sea'. In the second storm, of 13-14 July 1881, ten sixerns and 58 men were lost.

On 19 August 1848, more than 600 boats put to sea from Wick and 300 from Peterhead. The weather looked threatening in the evening, when the nets were shot. By midnight, the storm, a shrill south-easterly full of rain, had broken, and many boats ran

The wreck of the Oceanic *at St Monans, Fife, 1912.* SEA

for shelter. There was merely five feet (1.52m) of water at the entrance to Wick harbour and in the heavy seas no boat could enter. Twenty-five fishermen perished in the harbour mouth, and twelve farther out to sea, when their boats were swamped. At Peterhead, 30 fishing-boats were lost, 33 damaged or stranded, and 31 men died. Other, smaller ports along the coast also suffered, and in total 124 boats and 100 men were lost.

The most catastrophic storm in the history of Scottish fishing, remembered as the 'Eyemouth Disaster', occurred off the Berwickshire coast on 14 October 1881. The village barometer at Eyemouth was low that day, and the sky ominous; but the weather had been threatening for several days, and stocks of costly bait were begining to spoil. When one young man put to sea regardless, 281 men followed him. A few hours later the storm broke, and the boats ran desperately for harbour. Some were swamped or dashed on the rocks within sight of the helpless onlookers: wives, children and the elderly. From Eyemouth alone 129 men were lost; from Burnmouth 24; from Cove eleven, and from

Coldingham three. The Forth villages of Musselburgh and Newhaven lost 24 men. A hundred and seven widows and 351 fatherless children formed the apex of the suffering which that localized storm wrought.

12 Migrations

Throughout human history people have been on the move: on a large scale in times of war and famine, and on a small scale for the reasons that people nowadays move: to better themselves in their jobs, or to find jobs; to unite with a lover or spouse in his or her own community; to escape punishment or shame.

In St Monans, Fife, in the 1790s, there was a 'great decrease in fishing', and the fishermen were 'threatening an emigration to other places'. One fisherman and his family had already left and settled in Ayr. In Ayr itself, and adjacent Newton-upon-Ayr, in the 1830s, a migration of fishermen to Loch Fyneside reduced the fishing to a purely local industry.

To Ardrishaig came Laws, Bruces and Hamiltons, and to Tarbert Laws, Bruces and Hays. The Hay family merits special notice, owing to its having produced two significant figures in modern Scottish literature, the father and son John Macdougall Hay, author of the novel of Tarbert, *Gillespie*, published in 1914, and George Campbell Hay, linguist and poet, much of whose work took its inspiration from the Tarbert he remembered, and particularly the herring fishing and fishermen.

A crew of Shetlanders was brought to the Ross of Mull in 1789, on the initiative of the Duke of Argyll, to instruct the islanders in the catching and curing of white-fish. The farm of Creich was sub-divided into crofts to accommodate the natives, who were provided with free boats and lines; but the venture failed.

On the East Coast, fishing communities were in existence by the sixteenth century. Most of them were consolidated by the drift of farmworkers and others to the coast, but most new-founded fishing villages were colonized by fishermen from elsewhere. Thus, in Morayshire, Portknockie was founded by settlers from nearby Cullen in 1677; Findochty by Fraserburgh fishermen in 1716, and Portessie by Findhorn fishermen in 1727.

Throughout the nineteenth and into the twentieth century, there were massive seasonal migrations of fishermen from west to east, in search of engagements on board the drift-net boats which followed the herring shoals from the Northern Isles down the East Coast. Thousands of crofters, particularly from the Outer Isles, participated in these fisheries, and by the 1870s virtually the entire male populations of some districts joined the summer exodus. In Lewis, for example, 'only a few old men and boys' remained. The women, too, left in great numbers to find employment at gutting and packing (p37).

Before the coming of motor power and before decked boats became general, the movements of fishermen were often hampered by tide and weather, or limited by sheer distance. The custom therefore developed, particularly on the West Coast and in the Shetlands, where creeks and inlets abound, of using fishing stations when working from home. The practice, which had its parallel in the annual migrations of people and livestock to summer grazings, was an ancient one. In 1705, during the reign of Queen Anne, legislation was passed granting fishermen certain rights on the shores. That legislation was followed in 1775 by a more comprehensive Act which gave fishermen free use of the shores below the highest water-mark and of 100 yards (91.44m) of foreshore on waste or uncultivated ground, for landing nets and casks, etc.

The Campbeltown ring-net fishermen were accustomed, in the late nineteenth century, to camping in tents or living in wooden huts on the shores north of Carradale during the summer herring season. By night they fished in their open skiffs, selling their catches at sea to the buying-steamers, or *herring-screws*, which sped the fish to Glasgow in the morning. By day they cooked their meals and slept in the bays of their choice, which in time became associated with particular families.

Haaf-fishermen outside their lodges, Stenness, Shetland, about 1900. SEA

Caves were used as seasonal shelters by fishermen, as also by herds, vagrants and tinkers. Idrigill Cave, at the entrance to Loch Bracadale, Skye, was described in 1841 as a 'dwelling' of fishermen: 'They here hang up their nets to dry, cure their fish, cook their victuals, and sleep soundly on the dry sand with which part of the cave is strewn.'

The lobster-fishermen of Grimsay on North Uist used *bothies* (huts) on the Monach Isles, eight miles out in the Atlantic. The fishing season lasted from May until the end of August. The huts were thatched with *muran* (marram grass) and the fishermen's beds were simply sacks filled with the same grass, and with a blanket for cover. Peats were brought across from Grimsay as fuel.

The fishing stations of the Shetlands were both extensive - C A Goodlad has plotted no fewer than 98 of them - and well-organized. Each haaf crew built and maintained its own stone lodge, used year after year. At the start of the season, a roof of wooden laths and turf was put on and, when the season ended, the wood - 'so valuable in treeless Shetland' - was taken home.

Within the past half-century, numbers of East Coast fishermen have taken to basing their boats at such West Coast ports as Kinlochbervie and commuting by car, thereby preserving social ties with their native communities.

13 Community and custom

Fishing communities, until recently, were singularly close-knit. The sons of fishermen almost invariably themselves became fishermen and preferred to marry daughters of fishermen. This custom, however socially limiting, made good sense. Daughters of fishermen were reared to be wives of fishermen, and knew no other life. That life involved many hardships, which few outsiders could adapt to.

The following account, from Rathven in Banffshire, in 1842, states the matter very well:

> The boys go to sea as soon as they can be of any service to their fathers ... At eighteen years of age they become men, and, whenever they acquire the share of a boat, they marry, as it is a maxim with them 'that no man can be a fisher, and want a wife'. They marry, therefore, at an early age, and the object of their choice is always a fisherman's daughter, who is generally from eighteen to twenty-two years of age. These women lead a most laborious life, and frequently go from ten to twenty-five miles into the country, with a heavy load of fish ... They assist in all the labour connected with the boats on shore, and show great dexterity in baiting the hooks and arranging the lines.

Women also assisted in 'dragging the boats on the beach and in launching them', and 'sometimes, in frosty weather, and at unseasonable hours, [carried] their husbands on board, to keep them dry'. (Modern fishermen's wives have no such labours, but they still have the burden of raising a family and managing a household virtually single-handed.)

Intermarriage within these communities tended naturally to restrict the accession of new blood, and in many fishing villages a certain few families would become dominant by strength of numbers and inter-marital connections, or by their success. So it came about that particular names became associated with particular villages. Nicknames were, and remain, common in the fishing fraternity. Some bear obvious origins, in personal appearance or in a peculiar trait, while others are hereditary and obscure. But whatever their origin, nicknames were absolutely essential to the separation of individuals sharing names. In 1842, the minister of Cullen, the Rev George Henderson, listed the nicknames of three lots of Addisons there, Alexanders, Williams and Jameses. The Williams were 'Sheepie', 'Boatie-row', 'Calkinapin' and 'May's Wilsie'.

Fishermen are very superstitious and probably always have been, fishing being both physically dangerous and economically

Musselburgh fishwives dancing during the annual Fishermen's Walk, 1948.
C&F McKean, SEA

uncertain. The fisherman, therefore, has to guard his luck, and he was always vulnerable in the transitional stage between land and sea, that is, on his way to the boat. In that state of mental preparation for his time at sea there were certain persons and certain things he feared to meet, and so powerful was the belief in the malignity of the chance encounter with these unlucky influences that many fishermen would turn back, either to stay ashore or to perform some ritual of cleansing before setting out once more.

Women in general were to be avoided, especially red-haired women and old women with witchlike characteristics. Individuals, male or female, having some physical deformity or social irregularity - a tramp, for instance - might also be considered unlucky. In leaving harbour, a boat would never be turned *widdershins* (anti-clockwise), but 'with the sun'. At sea, too, the fisherman was always on his guard. To whistle at sea was to 'whistle up the wind', that is too much wind; and sticking a knife in the mast also raised wind.

There are many taboo words which cannot be spoken aboard a boat. These words vary in number and character from community to community, but the commonest are: minister, priest, rat, rabbit, salmon, pig and swan. There are, however, synonyms for all of these. A rat might be a 'long-tail' and a salmon a 'red fish', but these vary from locality to locality.

Many Hebridean women believed that it was unlucky to wear at sea clothes that had been dyed with the lichens called *crotal*, because crotal would go back to the rock whence it had come; and a woman must not comb her hair at night while her menfolk were at sea or else they could be drowned with their feet entangled in it. There was a widespread belief that herring were averse from human blood and would instantly desert a coast on which blood had been spilled in anger.

At Avoch, on the Moray Firth, when a fisherman was married, the best man would untie the shoe on the left foot of the bridegroom and form a cross with a nail or knife on the right-hand post of the church door, 'with a view, it is said, of setting at defiance the power of witchcraft'; and no marriage would be solemnized before twelve noon.

Religion and superstition each had its place in the fisherman's psyche, and if there was any conflict between the two it was seldom acknowledged. In the early nineteenth century, the herring crews of Latheron in Caithness engaged in worship after shooting the nets: 'On these occasions a portion of a psalm is sung, followed by a prayer, and the effect is represented as truly solemn and heart-stirring, as the melodious strains of Gaelic music, carried along the surface of the waters ... spread through the whole fleet.'

A street scene in Cromarty, 1880s. The tubs and creels in front of the houses are all connected with small-line fishing. The man with the can and spade has been digging lugworms for bait, and below the eaves of one of the cottages hangs a string of speldings, *or dried fish. These line-caught fish would have been too small for sale, so kept for home consumption.* SEA

The religious 'revival' which began in America in 1859 and reached Scotland in the following year, gripped the fishing communities along the Buchan coast. It was led by a Peterhead cooper and herring-curer, James Turner, who held meetings in his own curing-shed, lit by several oil-lamps and smelling strongly of 'salt fish and smoke'. He was reputed to have converted more than 8,000 persons along the coast in the two years before his death. Psalm-singing at sea was revived at that time, and the custom was started of 'kneeling down in the cabin for prayer together before they would let down a net'. That revival was followed by another, which took hold in 1921 during the East Anglian fishing, at a time when many skippers and crews were deeply in debt and troubled in mind.

14 Daily lives

The typical fisherman's cottage in the early nineteenth century, and farther back, would have differed little from that of his country counterpart. It was generally a single room with small windows. Inside, an earthen floor, simple furniture, with usually a box-bed in a recess, and the whole interior illumined by fish-oil lamps. From the smoke-blackened rafters would hang the paraphernalia of his occupation: nets, buoys, ropes, etc., and the ubiquitous dried fish. If he was a crofter as well - as were many fishermen, particularly in the Northern and Western Isles and the West Coast - the kitchen end of the house would probably be occupied by fattening livestock, perhaps a pig, a calf, and half-a-dozen lambs, in addition to his family. The fire would have burned in the middle of the floor, with a heavy pot-chain hanging over it, and the smoke - or some of it - going out through holes in the thatched roof.

Some cottages, designed purposely for fishermen in the later nineteenth century, had net-lofts under the roof-pitch, and small doors set in the gables for hauling gear in and out of storage. In the early nineteenth century, the magistrates of Aberdeen planned two-storied houses to replace the old dwellings at Footdee and Torry, but the fishermen rather perversely refused to live upstairs, arguing that 'it would have been very inconvenient ... to lug their long lines and their heavy baskets up stairs'. They also refused to give up their earthen floors, on the ground that it would have been 'next to impossible ... to have kept a wooden floor clean'.

In the open boats there were generally no cooking facilities at sea (though in some, a brazier was carried). The fishermen's diet was therefore simple: fundamentally oatcake and water, though the Shetland haaf-fishermen might carry *bland* (sour whey). Sometimes a tasty accompaniment would be available on the fishing grounds. The Loch Fyne fishermen ate *bonnach èineachan*, raw milt of herring between oatcakes, and Lewis fishermen had *ceapaire*, which was cod's liver with oatcake.

When stoves were fitted into the boats, fish itself became the staple diet, when available. Herring-fishermen would have herring, boiled or fried, thrice daily, accompanied by oatcakes or

Breakfast in the forecastle of the ring-netter Kittiwake *of Campbeltown, about 1950. From left to right, Skipperowner John Wareham, Willie Scally, and Archibald 'Scone' Black.*

potatoes, if they had them. On warm summer days, a batch of herring might be split open, salted and peppered liberally, and spread to toast in the sun. Seabirds, principally guillemots, shags and cormorants, were shot at sea to supply meat. A quick and tasty meal, known as *skirlie*, could be served by frying oatmeal and onions in fat. *Crappit heids* was a traditional dish among fishermen: fish livers and meal packed into a cod or haddock head (or stomach) and boiled with fish flesh.

Hard sea-biscuits - baked with flour, water, sugar, salt and, optionally, lard - were carried to sea by the stone. These biscuits were a valuable stand-by if bread ran short. Tea, sugar and salt were essential provisions. Salt beef was carried on some East Coast drifters, to last the season. By the early twentieth century, such relative luxuries as bread, butter, jam, treacle and tinned condensed milk for tea were brought aboard, and in some boats tobacco was supplied to the crews as part of the stores. Provisions were paid out of general expenses. When the author and illustrator Peter F Anson joined the crew of a Moray Firth drifter for a trip to the West Coast in the 1920s, he was delighted with the standard of cuisine, enthusing over the fried herring 'that any first class chef would have been proud of', and the 'light and tasty' plumduff.

Fishermen, on the whole, were a strong and healthy breed, but there were certain ailments to which they were particularly prone. Piles was a common condition, and the high incidence of fistula among fishermen was explained, in 1849, in terms of 'the difficulty they experience in relieving nature while at sea in their open boats'. Rheumatism was also common. In summer, fishermen hauling the nets were frequently subjected to irritation of skin and eyes from the stinging properties of certain jellyfish, chiefly the lion's mane, known in Scots as the *scalder*. Salt-water boils on the hands affected some fishermen.

The dress of fishermen changed from generation to generation according to fashion, but one constant was the dark-blue oiled-wool jersey, knitted in the round, i.e. without any sewing, and decorated with a variety of symbols which varied from community to community. These jerseys were knitted, with great skill, by the womenfolk, who maintained their industry even when out walking. There was a wonderful range of headwear: tall silk hats worn ashore and, at sea, sealskin caps, high soft hats, cheese-cutters, blue bonnets with tassels on top, glazed straw-hats, flat cloth caps in the early twentieth century and, latterly, close-fitting woollen hats. The buss fishermen of the late eighteenth century worked in a 'warm great coat', boots and 'skin aprons'. Later, oilskin suits came into use. These were preserved by coating with linseed-oil, as were the heavy thighlength leather sea-boots, which were universal until replaced by rubber boots in the 1920s.

15 The modern age

The modern age of fishing, as of every other traditional industry, came in with mechanical power. The first form of power was steam, followed by oil, and then by electricity, which is generated now by diesel-driven engines, and on which runs the navigational and fish-finding equipment encircling the modern skipper in his weather-proof wheelhouse.

This 'progress' which has taken place in little over a century has utterly transformed the lives of fishermen. The skipper, in the old sailing boats, had his place at the stern. He sat there, exposed in all weathers, managing the helm, directing fishing operations and involving himself in the basic work of the boat. The modern skipper, especially on the bigger class of boats, is confined to his wheelhouse and seldom sets foot on deck. He can put on carpet-slippers and sit back on his revolving chair; but his is a lonely existence, and during net-towing, conversational needs are more likely to be satisfied by radio contact with other skippers than by his own crew. And far from communing with nature, most of his time is taken up with scanning his instruments. He is also likely to be over-weight, and if a smoker he will almost certainly be a chain-smoker.

The crewmen, for their part, have virtually nothing to do with decision-making. Their senses, once so valued, have become almost redundant. Muscle-power, too, is no longer the attribute it was. A lifetime spent pulling oars and halyards and hauling nets by the power of their arms alone, brought men to a certain degree of fitness (though confinement aboard a boat for long periods has never been conducive to general fitness). The engine now does all the work of powering the boat, and the net itself is taken aboard by power-block. The deck-hands too mostly function out of the weather, in deck-shelters, where they sit and process the fish while the sea rolls by and the sky shifts its eternal cloud patterns above them, unseen except in glimpses. Against all that, the modern fisherman works in weather that his forebears would not have considered venturing out in. It is significant, too, that developments in fishing gear and in the various appliances that support the fishing effort no longer originate with the fishermen themselves, but are the products of landbased specialists for the most part.

Steam-power came first to the trawling industry. Trawling, by its very nature, benefited immediately: the greater the speed at which the net could be towed across the seabed, the more ground could be covered, and therefore the more fish caught. Steam-trawling arrived in Aberdeen in 1882-83 and the fleet, which became the biggest of its kind in Scotland, grew steadily until in 1938 it numbered 255 vessels. The port itself grew apace. Shipbuilding revived: between 1883 and 1903, 267 trawlers were built.

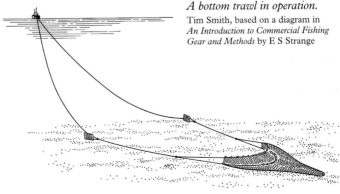

A bottom trawl in operation.

Tim Smith, based on a diagram in *An Introduction to Commercial Fishing Gear and Methods* by E S Strange

Daily auction sales were introduced to dispose quickly of the fish, which were sent to London on special trains. Ancillary industries, from box-making to ice-making, sprang up.

The fishing range of the Aberdeen trawlers was extensive: the North Sea, Orkneys and Shetlands, and around Fair Isle; west of the Hebrides and around St Kilda; Iceland; Skagerrak; Lofoten. In time, skippers were tempted to extend their explorations still further as the customary grounds became over-exploited. In 1929 an Aberdeen trawler ventured for the first time to the Barents Sea, a round-trip of 3,700 miles, and in the following year Bear Island was prospected unsuccessfully.

The second most important port in Scotland was Granton, on the Firth of Forth. Its fleet - by 1925, 85 vessels - operated mainly in near waters: the North Sea and around the Shetlands and Fair Isle. Dundee was the third-ranking port, with a fleet seldom exceeding ten vessels and working mainly around May Island and Bell Rock. In the post-war years the fleet dwindled until, in 1954, trawling ended there. In 1899, for the first time, the total catch of trawled white-fish exceeded that of line-caught fish.

Shooting a seine-net. The vessel has set the net and is
returning to the dan to commence hauling in the ropes.
Tim Smith, based on a diagram in *An Introduction to Commercial*
Fishing Gear and Methods in Scotland by E S Strange

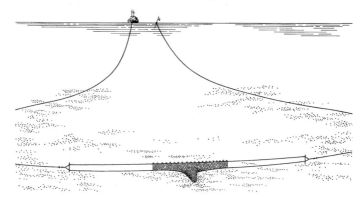

Of all the classes of fishermen, the trawlermen undoubtedly led the hardest lives. Deckhands worked on lurching open decks, often awash with driving spray and, in winter, coated with ice. The work was monotonous and continued day and night for weeks at a stretch. Since the trawler crews were essentially wage-earners, unlike their share-fishing counterparts on the inshore boats, they had little commitment to or pride in their boats and their calling.

In the 1920s, a new method of white-fishing was adopted by Scottish fishermen. This was the Danish seine-net, so named because pioneered in Denmark in the mid-nineteenth century. Though technically a seine-net, in effect the long ropes of the gear did the fishing. The method developed initially in the Moray Firth, chiefly at Lossiemouth, and spread gradually around the Scottish coasts. In 1950, for the first time, the seine-net fleet out-fished the trawl fleet: 1,604,490 hundredweights against 1,469,793 hundredweights. By then, many East Coast boats, notably from Macduff and Lossiemouth, were spending much of their year fishing on the West Coast, landing at Kinlochbervie and Oban and consigning catches to market by road or rail. The typical seine-net boat was under 70ft (21.34m) and equipped to fish seasonally with drift-nets.

Steam-drifters arrived in Scotland in 1898, and by 1912 there were more than 800 of them operating, almost exclusively East Coast-owned. The typical steam-drifter, common to England and Scotland, was about 90ft (27.43m) in length with a beam of 19ft (5.79m). Straight of stem and sweeping low to the stern, she was a graceful craft and could make about nine knots speed. The steamer could work, wind or no wind, pursue herring at greater distances from shore, and land catches ahead of the older boats in calms.

There was no immediate abandonment of the fifie and the Zulu. They were much cheaper to buy and to run and, besides, motor-power was on its way. Experiments with motorpower began in earnest in 1905 with the *Pioneer*, but that and subsequent trials were unsatisfactory. The first commercially successful motor vessel was the 65ft (19.81m) Eyemouth fifie, the *Maggie Jane*,

Seine-netters berthed at Macduff, Banffshire, in 1964. The crewman standing left is splicing a rope. In the foreground is a seine-net winch, which coiled mechanically the miles of rope used in the fishing operation. SEA

fitted in 1907 with a 55hp Gardner engine. The first West Coast boat to be converted to motor-power was the ring-net skiff *Brothers* of Campbeltown. She was fitted with a 7-9 h.p. Kelvin in 1907, on the initiative of her skipper-owner, Robert Robertson (p32)

By 1914 there were 239 motor fishing boats in the Firth of Clyde, and 56 elsewhere on the West Coast, with 361 on the East Coast and 38 in the Orkneys and Shetlands. In that year there were no fewer than 50 models of marine engine on the British market. The revolution was well under way. Sail fishing boats lasted, however, until 1966. There were eleven registered in that year, and none in the following year. Ironically, the sailing-boat outlasted the steam-drifter.

The invasion of the Shetland herring grounds in 1965 by a fleet of more than 150 Norwegian purse-seiners caused widespread alarm among Scottish fishermen, who saw the visitors taking hundreds of tons of herring with each set of the net, against their own 40 or 50 crans. They also saw that the pursers could work night and day until loaded with herring. That these catches were destined not for human consumption, but for Norwegian fish-meal factories, to be reduced into animal food, intensified the disquiet. There were appeals for the prohibition of purse-seining in Scottish waters, but these came to nothing, and in the summer of 1966 the first Scottish purse-seiner was fitted out on the initiative of Skipper Donald Anderson of Peterhead. She was the 77ft (23.47m) *Glenugie III*, and her success encouraged other Scottish fishermen to convert to purse-seining. In 1967 there were six Scottish pursers at work and in 1969 sixteen, including partner vessels.

The purse-net is a close relative of the Scottish ring-net. The one was pioneered in the fjords of Norway, the other in Scottish sea-lochs. Purse-seines are designed on a scale hitherto unprecedented - up to 4,000ft (1,200m) long and 800ft (240m) deep - thanks to technological innovations, chiefly the invention of the net-hauling power-block and the availability of light synthetic fibres, which reduced the bulk of netting and gave it extra strength. The result was a rapid and alarming increase in catch rates.

Moray Firth steam-drifter the Mistress Isa *heading out to the fishing grounds.* SEA

Mid-water trawling - the towing of a light trawl-net, between surface and seabed - developed apace, and was firmly established by 1970. Mid-water trawling is operated mainly by two boats in partnership, and was a cheaper option than purse-seining, being open to boats of a smaller class. By 1969, only fifteen drifters and 52 ringers remained in the Scottish fleet; by 1972, the great bulk of the Scottish herring catch was taken by pair-trawlers and purse-seiners.

Modern fishing is bedevilled by restrictions, increasingly dictated by European Community policy-making in Brussels. The position of Scottish fishermen in these political cross-currents is a complex one, but there is a course ahead, rough though it threatens to be. Basically, there are no longer any state inducements for fishermen to continue fishing, as in the recent past, when grants and loans were available. On the contrary, fishermen are being encouraged to stop fishing through the Government's 'decommissioning' scheme, which aims to reduce catching capacity by paying boat-owners to scrap their vessels.

It seems likely that the inshore fleet will continue to diminish, while the biggest class of boats - using purse-seines, Danish seines and the various trawls - holds on. Indeed, investment continues in that section of the fleet, an example being the 243ft (74.20m) pelagic 'tank' trawler *Altaire*, reckoned to be the biggest vessel of her kind in the world. She was designed and built in Norway in 1994, but not for owners in Aberdeen or Peterhead or Fraserburgh, the three major ports in Scotland. She is owned by Skipper John Peter Duncan of Northmavine in Shetland, along with seven other fishermen, and represents an investment, gear included, of close on £10 million.

In such communities as the Shetlands and the North-East, fishing is more than just a job. It is a tradition and an economic necessity. Communities such as these will keep fishing alive at all costs, because if fishing dies the communities themselves must also die.

FURTHER READING

ANSON, P F *Fishing Boats and Fisher Folk*, London 1930.

ANSON, P F *Fishermen and Fishing Ways*, London 1932.

BOCHEL, M 'The Fisher Lassies', in *Odyssey: The Second Collection* ed Billy Kay, Edinburgh 1982.

BUCHAN, A R *The Peterhead Whaling Trade*, Peterhead 1993.

DYSON, J *Business in Great Waters*, London 1977.

GOODLAD, C A *Shetland Fishing Saga*, Lerwick 1971.

GRAY, M *The Fishing Industries of Scotland, 1790-1914*, Oxford 1978.

HARDY, A *Fish and Fisheries*, London 1959.

HODGSON, W C *The Herring and its Fishery*, London 1957.

HUNTER, J *The Making of the Crofting Community*, Edinburgh 1976.

MARTIN, A *The Ring-Net Fishermen*, Edinburgh 1981.

MARTIN, M *A Description of the Western Islands of Scotland*, 1703.

MILLER, H *Tales and Sketches*, Edinburgh 1889.

NETBOY, A *The Atlantic Salmon*, London 1968.

New Statistical Account of Scotland 15 vols, 1845.

PENNANT, T *A Tour in Scotland and Voyage to the Hebrides*, 1772.

SMITH, P *The Lammas Drave and the Winter Herrin'*, Edinburgh 1985.

SMITH, R *The Whale Hunters*, Edinburgh 1993.

Statistical Account of Scotland 21 vols, 1791-9.